PENGUIN BOOKS

Don't Swallow Your Gum

Don't Swallow Your Gum

And Other Medical Myths Debunked

DR AARON CARROLL AND
DR RACHEL VREEMAN

PENGUIN BOOKS

PENGUIN BOOKS

Published by the Penguin Group
Penguin Books Ltd, 80 Strand, London WC2R ORL, England
Penguin Group (USA) Inc., 375 Hudson Street, New York, New York 10014, USA
Penguin Group (Canada), 90 Eglinton Avenue East, Suite 700, Toronto, Ontario, Canada M4P 2Y3
(a division of Pearson Penguin Canada Inc.)
Penguin Ireland, 25 St Stephen's Green, Dublin 2, Ireland (a division of Penguin Books Ltd)
Penguin Group (Australia), 250 Camberwell Road, Camberwell, Victoria 3124, Australia
(a division of Pearson Australia Group Pty Ltd)
Penguin Books India Pvt Ltd, 11 Community Centre, Panchsheel Park, New Delhi – 110 017, India
Penguin Group (NZ), 67 Apollo Drive, Rosedale, North Shore 0632, New Zealand
(a division of Pearson New Zealand Ltd)
Penguin Books (South Africa) (Pty) Ltd, 24 Sturdee Avenue, Rosebank, Johannesburg 2196, South Africa

Penguin Books Ltd, Registered Offices: 80 Strand, London WC2R ORL, England

www.penguin.com

First published in the USA by St Martin's Press, 2009
Published in Penguin Books 2009
1
Copyright © Dr Aaron Carroll and Dr Rachel Vreeman
All rights reserved

The moral right of the authors has been asserted

Typeset in 11/13pt Bembo by Palimpsest Book Production Limited, Grangemouth, Stirlingshire
Printed in England by Clays Ltd, St Ives plc

ISBN: 978-0-141-04336-4

Penguin Books is committed to a sustainable future
for our business, our readers and our planet.
The book in your hands is made from paper
certified by the Forest Stewardship Council.

For Aimee, who actually believes that there is nothing I cannot do, and Joe, L of my L, who wanted to get us myth-busting capes

Contents

Part VI: 'Vaccines made my baby autistic' *Myths That Spark Controversy and Debate*

Introduction

Is wearing a hat the key to staying warm when the temperature plummets? Do you wish you could use more than 10 per cent of your brain? Have you switched to waxing your legs because shaving made the hair grow back faster and thicker? Is it a struggle for you to drink the recommended eight glasses of water a day?

Every day, you hear or think things about your body and health that are not true. Maybe these ideas are simply unproven. Or maybe these ideas about your body and how to keep it healthy have actually been shown scientifically to be false.

And yet we still see these things on TV, read them in magazines or hear them from our friends. Even your doctor may have told you one of these myths.

One of the dirty little secrets in the practice of medicine is how little of what we doctors do is actually proven. People assume that if doctors recommend something then it must be correct. To *know* something is true, however, requires scientific research, and good research requires time and money. Although there are millions of people and billions of pounds invested in scientific research, there just isn't enough to go around to answer every question; especially since the focus of modern research is usually on the most serious problems and the most advanced drugs and procedures.

Often when a doctor tells you to do something, it's just his best guess. And that's okay. A good doctor builds on his training, experience and knowledge to give you his best advice, so most of the time this medical advice will be useful and helpful. However, another doctor may give you different advice. Who's right? How do you know whom to trust?

If this is happening with important medical issues, what about those things that are less critical – the things your mother warned you about? Or the things your friends told you they saw on TV, or even the stuff you read in popular health books?

The fact is that often we just don't know what's true. That's where the idea for this book came from. Many things you believe about your health, things you were told as a child, are simply unproven. Again, that's okay, but these ideas should not be given the same weight and credence as those that are proven. It's in our best interest to understand where these unproven beliefs came from, and then judge for ourselves whether they are useful.

Some of these myths have actually been studied. So we can look at what the studies say and make a judgement. How do we decide whether to trust scientific studies? We are both doctors and researchers. In our professional lives, we spend a great deal of our time teaching people, from parents to other doctors, how to understand health research. As researchers, we strive to maintain a sense of fairness in our work; we are careful not to make any decisions before the experiments are complete. Therefore we are able to accept whatever the science tells us, regardless of what we might have believed before. We want to give you a crash course in health research and scientific studies so that you can understand how science helps us decide whether a belief is right or wrong. As part of this crash course, we suggest a quick look at the section Boring Research Terms That You Might See in This Book. We don't want you to believe something is a myth just because *we* said so. We want you to understand *why* we said so.

The best possible type of study is what the scientific world calls a 'randomized controlled trial' (see the Boring Terms section.) In these studies, people are secretly given one of two or more treatments. In the best research, no one involved in the study knows which person is getting what treatment. By looking at what happens to people in this kind of study, we can tell what effect the treatment or situation *caused*. Randomized controlled trials are the only way to prove causation. You should be sceptical when anyone tells you something has been 'proven to work' or 'proven to cause' something unless the outcome was a result of a randomized controlled trial. But these types of trials are pretty rare in the medical world because they are expensive and complicated. Sometimes they are even unethical. You have probably heard tobacco companies say that smoking has never been proven to cause cancer. This is because there has never been a randomized controlled trial of smoking and cancer. And there never will be. Can you imagine anyone approving a study that secretly forced some people to smoke so that we could see if they got cancer? That would be crazy! And wrong!

When we simply can't get a randomized controlled trial, we have to look at the next best thing. Through other types of studies, called epidemiological or cohort studies, we can look at 'associations'. An association is a relationship one thing has with something else. Association and causation are not the same. While we can't prove that smoking *causes* cancer, there is an overwhelming amount of evidence that smoking is *associated* with cancer. With evidence that reveals an association between smoking and cancer, we can't say we are 100 per cent sure that smoking causes cancer, but we are as close to being sure as we are ever going to get. The scientific studies that tell us about associations usually involve big groups of people (the bigger, the better), where we can best see if these people have certain things in common (like smoking and getting cancer).

Many of the myths in this book have this type of evidence stacked against them. Maybe there aren't any randomized controlled trials to dispel the myth, but there are large epidemiological or cohort studies that point to that answer. When scientific evidence for or against something builds up, we argue, you should believe the science. In this book, you will always see us argue on the side of science.

Keep something else in mind as you read this book: that you

cannot prove a negative. Although we can tell you that something has never, ever happened in the history of the world, we
cannot offer definitive proof that this thing won't *ever* happen.
This does not mean you should *expect* it to happen. It is not
logical to believe that something is true or that something is
going to occur only because there is no absolute proof that it
is false.

For instance, in the history of the world, no one has ever been
born who could fly. We can't prove, or say with 100 per cent
certainty, that someone won't be born tomorrow who can, but
it is really, really, really unlikely. So it's okay for us to say, even
without absolute proof, that people can't fly.

In this book, we will examine a lot of beliefs about your body
and your health. We will lay out the science as best we can, based
on everything we can find in medical and scientific literature. We
will argue that you should decide what to believe based on the
evidence or lack thereof.

What really concerns us are those myths which great randomized controlled trials have already disproved and which people still believe. This is frustrating, because the jury is in — there
won't and shouldn't be any more studies. All the research indi-
cates that the myth is untrue, but people just don't want to accept
that.

We know that people don't like to hear that they are wrong
(Aaron especially — he always thinks he is right). Discovering
that something you believed is not true can be disturbing and
unsettling. When we published the first of these myths in the
British Medical Journal in December 2007, we were shocked at the
strong reactions it provoked. Some people just can't let a myth
go.

Some of you will read this book and still refuse to accept what
the studies show you to be true (or false). We have provided
extensive references for those of you inclined to investigate further. You may be surprised not only by the scarcity of evidence

supporting some of your beliefs but also by the volume of evidence disproving others. We tried to show you everything; we included all the evidence we could and told you when there was none.

Keep an open mind. More often than not, this book will make your life easier. Moreover, it will give you a number of smart answers for your mother.

Boring Research Terms That You Might See in This Book

- **Association** – when one thing is shown to be linked to another, but is not necessarily the cause. This is not the same as **causation**.
- **Biases** – factors in studies that may make the results skewed in one way or another. To avoid biases, researchers will try to make them **randomized**, **blinded** and **placebo-controlled**.
- **Blinded trial** – a study in which subjects and/or researchers do not know which treatment subjects are getting until after the trial is complete.
- **Case-controlled trial** – a study that uses **matching** to compare people with and without certain conditions or characteristics to see if they have an **association**.
- **Causation** – when one thing has absolutely, positively been proven to be the reason that something else happens. Can only be done with a **randomized controlled trial**.
- **Clinical significance** – a subset of results with **statistical significance** that also have meaning in real life.
- **Cochrane Collaboration** – a public warehouse of high-quality **systematic reviews** and **meta-analyses**.
- **Cohort study** – a study that usually follows large groups of people over time to see what happens to them, without interfering in their care.
- **Matching** – in a **case-controlled trial**, different types of people are paired with each other based on pre-selected characteristics, like age or sex.
- **Meta-analysis** – a special type of **systematic review** that uses statistics to combine the results of all the included studies.
- **Placebo-controlled** – well-designed studies have one group

getting a 'fake' or placebo treatment in order to hide which people are getting the intervention and which are getting nothing.

- **Randomized** – in the best kind of studies, people are chosen at random as to how they are treated. Such studies are called randomized trials.
- **Randomized controlled trial** – a study that is **randomized** and **placebo-controlled**. These studies are the only type that can prove **causation**.
- **Statistical significance** – a mathematical calculation showing an association or causation to be very probably true. Differs from **clinical significance** in that the significance may not be meaningful in real life.
- **Systematic review** – A formalized compilation, using a documented scientific method, of all the relevant studies on a particular topic.

PART I

'Look at the size of his feet!'

Myths About Your Body

Myth: Men with big feet have bigger penises

Have you ever noticed a man with particularly big feet and wondered whether other parts of him were just as large? While some claim that a man's penis size can be predicted by the size of his feet, others say that it's the size of his hands or even his nose that really gives away the secret of what's in his pants. This idea of comparing body parts to estimate hidden assets may have originated with discriminating shoppers, but it may also have its roots in real science. The Hox gene in mammals plays a role in the development of the toes and fingers, as well as the penis or clitoris. Given how rarely one hears 'gene expression' mentioned along with talk of penises and feet, it seems much more likely that this myth springs from our human desire to identify patterns – even when the pattern is not really there. We like to have explanations for things we see, and we like to group like things with like things (in this case, appendages on men).

Despite the similar genetic controls for these protuberances, men with big feet do not necessarily have bigger penises. While the prospects for estimating penile length with a quick glance at a man's feet seemed promising at first, science now shows us that this is not the case. A study of 60 men in Canada suggested that there was a weak but statistically significant relationship between penile length and both body height and foot length (remember, statistically significant doesn't necessarily mean significant in real life). However, a larger study of 104 men, done by two urologists, Drs Shah and Christopher, which measured penises when pulled to their longest lengths, found that shoe size and penis size were not correlated with one another. An even larger study called the 'Definitive Penis Size Survey', which looked at 3,100 men, also found no relationship between shoe size and erect penis size. This was even the case when men reported their size, as opposed to direct measurement. And when men are asked to report the length of their own members, well, let's just say that science also shows us that exaggeration routinely comes into play. Furthermore, the 'Definitive Penis Size Survey' was never peer-reviewed or published in a scientific journal. Therefore we trust the results from Drs Shah and Christopher the most, but this survey backs them up. Studies also show no link between finger length and penis size. You can look at a guy's feet or hands all you want, but they won't tell you anything about how he measures up elsewhere.

Myth: You use only 10 per cent of your brain

You know that you can achieve your dreams if you just put your mind to it. After all, you're using only 10 per cent of your brain, right? Imagine what you could do with that other 90 per cent!

It's time for a reality check. People have believed that we use only 10 per cent of our brains for over a hundred years. Unfortunately, all that means is that people have been wrong for over a hundred years. As early as 1907, self-improvement gurus and motivational speakers were convincing their audiences that they could reach greater heights of achievement if only they could tap into some of their unused latent brain power. Some people even claim that Albert Einstein first said that most people use only 10 per cent of their brains, or that he was a genius because he used more of his brain than the rest of us. Neither of these claims is true! There is no official record of Einstein saying such a thing.

The myth of the unused brain has been debunked in great detail by an expert in neuroscience, Dr Barry Beyerstein. Many studies of patients with brain damage suggest that harm to almost any area of the brain has specific and lasting effects on a human being's capabilities. If this myth was true, it would not be a big deal to hurt various parts of your brain. Most of the time, however, that's not true at all. You will be affected by damage to almost any part of your brain.

Different types of brain-imaging, including CT scans, MRI scans, and even more detailed techniques, show that no area of the brain is completely silent or inactive. Much more than 10 per cent of the brain is busy at work virtually all the time. Furthermore, the many functions of the brain are localized to very specific areas of the brain. Each region has its own special job. When brain surgeons go in and probe the brain, area by area, they can't find the 'non-functioning' 90 per cent, because they see functions for almost every area. Moreover, when scientists observe the responses of individual brain cells or neurons (called 'micro-level localization') they do not find any gaps or inactive areas. Even studies of cell metabolism, which look at how the parts of the brain metabolize or process chemicals, reveal no dormant areas.

As depressing as this may be, you are probably stuck with what you've got. You are, in fact, using 100 per cent of your brain. Of course, you can still question the old adage 'you can't teach an old dog new tricks', since you *can* always keep learning new things. In fact, many studies show that keeping your brain stimulated might help to decrease dementia and brain impairment as you age.

Myth: Your hair and fingernails continue to grow after you die

Have you ever walked across a cemetery and thought about the growing, curling fingernails of the corpses beneath? This disturbing idea is probably the kind of thing you heard around a campfire as a kid. It has such a morbid appeal that artists have used this image in books and movies for a long time. Johnny Carson even joked, 'For three days after death, hair and fingernails continue to grow, but phone calls taper off.' Despite the popularity of this idea, it's just not true. To quote the expert opinion of forensic anthropologist William R. Maples, 'It is a powerful, disturbing image, but it is pure moonshine. No such thing occurs.'

This myth does have some basis in reality. After you die, your body dries out, or becomes dehydrated. As the skin dries out, it shrinks (this is a different kind of shrinkage than what happens to a man in cold water). The shrinking or retracting of the skin around the hair and nails makes them look longer or more prominent compared to the shrunken skin. It's just an optical illusion though; the nails and hair haven't actually grown at all. In order to keep growing, hair and nails require a complex mixture of hormones that are just not available to the body after death. Studies of the cellular regulation of hair growth confirm that a person must be alive for their hair to keep growing. There's no need to book a haircut and manicure for your corpse – no matter how long it's been since the last one.

Myth: If you shave your hair, it will grow back faster, darker and thicker

If you are a woman with a little more hair on your upper lip than you would like, you've probably always thought that shaving that hair would make it grow back darker and thicker than before. Your mother probably warned you the first time you wanted to shave your legs that the hair would only return – but worse. And even informative health websites state that shaving makes the hair grow back darker and thicker. Those of you who have shaved some part of your body have witnessed how quickly that dark stubble seems to pop up.

There are great scientific studies that prove that the hair you shave off does not grow back any darker or thicker than it ever was. As early as 1928, a clinical trial demonstrated that shaving had no effect on hair growth. When the researchers in that trial shaved patches of hair on some people but not on others, they did not find any difference in how fast the hair grew back. More recent studies confirm these findings.

The key to understanding this myth is knowing what really happens when hair is shaved off. Shaving removes the dead portion of hair, not the living section beneath the skin's surface. As it doesn't touch the portion of the hair that is responsible for growth, it is unlikely that shaving could change how fast the hair grows or what it looks like. In contrast, waxing and other forms of hair removal that pull the hairs out from below the skin actually *can* alter how fast the hair grows back. In fact, these methods, not shaving, might push the hairs into a phase of more rapid growth. So why does this myth persist?

An optical illusion is probably to blame. When you slice off the hair with a razor, it leaves a sharp end. Because these shaved hairs lack the tapered look of unshaven hair, it appears that the hair itself is thicker (even though it's not). It also may have a rougher feel because of that sharp edge. Additionally, the new hairs growing in have not yet had a chance to be lightened by the sun or other chemical exposures, so they are initially darker than existing hair – though they will lighten up just like the other hairs over time.

And think about it – if this myth were true, we would have a way to forestall baldness in men. And, as Aaron so often tells his friends, shaving your head will absolutely not make your hair grow back thicker and faster!

Myth: If you pull out a grey hair, two grow back in its place

As people get older, their hairs begin to turn grey. Plucking out those first grey hairs may seem like a good option, but many people worry that even more will sprout to replace them. The truth is your grey hairs *will* multiply as time goes on, but the plucking has nothing to do with it. Each hair grows out of a single follicle. Pulling one out is not going to make two hairs grow out of that one follicle. Furthermore, when you pull a hair out, it only grows back at a rate of 1 cm per month. During those months that the original grey hair is growing back, it is normal for other hairs around it to turn grey on their own. The idea that plucking produces more grey hairs is a myth. The plucking had nothing to do with the increase; time did.

Myth: You'll ruin your eyesight if you read in the dark

'Stop reading in the dark or you'll ruin your eyes!' You may remember hearing this from your parents when you were curled up in bed as a kid with your torch and the book that you just couldn't put down. Now, when you see other people or maybe even your own children reading away in dim light, you want to flip the light switch on or scold them in just the same way.

Dim light can certainly make you have difficulty focusing. It can also decrease how often you blink, making you uncomfortable because your eyes get dry and you squint for too long. However, the bottom line is that the effects of eye strain do not last. Once you return to good lighting, the effects go away.

There is simply no evidence proving that reading in the dark will ruin your eyesight for ever. In the face of no clear scientific evidence, we have to look at what other sources we can find – expert opinions, related studies and historical trends. The majority of ophthalmologists conclude that reading in dim light does not damage your eyes. Although reading in dim light can cause eye strain, with multiple temporary, negative effects, it is unlikely to cause a permanent change in the function or structure of the eyes.

One study did examine how the rate of blinking decreases during intense reading for patients with disorders that cause dry eyes, such as Sjogren's Syndrome. For patients with this syndrome, the decreased blinking and eye strain during reading can result in a temporary decrease in how well they can see. However, even in people with this condition, visual acuity improved when the patients stopped reading, again suggesting that the eyes return to their normal baseline when the strain is removed.

On the other hand, one review article on short-sightedness does conclude that 'increased visual experiences', such as reading in dim light or holding books too close to the face, could result in 'impaired ocular growth and refractive error' (in other words, reading in dim light might ruin your eyes). The primary evidence cited to support this claim is that short-sightedness is becoming more and more common, and that people who read more are more likely to be short-sighted. The author notes that this hypothesis is just beginning to 'gain scientific credence'.

In examining this argument, we must consider several important facts. First of all, association is not the same as causation. Just because more people who read a lot are short-sighted does not mean that the reading in dim light causes their short-sightedness. Even if the two are linked, the key factor may be the amount a person reads, not the amount of light present where the reading takes place. Another factor to consider is historical trends in lighting. Before the invention and widespread use of light bulbs, people had to rely on reading by candlelight in dark rooms. Now, most of us have access to light for reading whenever necessary. We have never had better light for reading in the history of the world. In that sense, the fact that more people are short-sighted today, when the world is so well lit, does not support the idea that reading in dim light hurts your eyes.

Thus our conclusion is that definitive scientific data do not exist to support or refute the claim that reading in dim light will ruin your eyesight, but the majority of experts believe (and common sense suggests) this is not true.

Myth: If you don't shut your eyes when you sneeze, your eyeball will pop out

You know what happens when you sneeze. The feeling in your nose builds up, your eyes close, and the sneeze explodes out. And thank goodness your eyes close, because otherwise your eyeballs would pop right out of their sockets with the pressure. Right? This idea has been known to scare some of us while driving. Do we shut our eyes and risk a crash, or do we leave them open and risk our eyeballs popping out? Neither option seems very appealing.

Well, we have to call this one a half-truth. It is physiologically possible for your eyeball to come out of the socket when you sneeze. Back in 1882, the *New York Times* tells us that an unfortunate woman in Indianapolis actually had her eyeball explode after she sneezed while on a streetcar. The article describes her as left in excruciating pain, needless to say. In the current medical literature, we could not locate any documented cases of eyeballs subluxing, or popping out, after sneezing. On the other hand, vomiting hard and frequently can make your eyeball pop out. In very rare cases, an eyeball can pop out spontaneously without anything else happening; the people that this happens to usually have weird muscles and can actually learn to pop the eyeball out voluntarily. Certain types of trauma to the eye can also cause it to pop out. But, for sneezing, we only have the sad, nineteenth-century story of the lady on the streetcar in our own town.

We are left with the expert opinion that it is possible but incredibly rare for an eyeball to pop out when you sneeze. Even if this were to occur, closing your eyelids would be unlikely to prevent the eyeball from coming out. A few years ago, the American television show *Mythbusters* forced someone to sneeze with their eyes open and noted that the person's eyeballs did not pop out. This is just one anecdote and is not good enough evidence to prove that it is not possible. But it does help to remind us that it would be extremely unlikely and that, despite all the horror stories, the eyes are very, very, very secure in their sockets.

Myth: The average person swallows eight spiders a year

Rachel hates spiders. This situation was not improved when an East African jumping spider bit her last year. Even though Rachel knows all of the logical reasons why she should like spiders, they really creep her out. And Aaron had a horrible experience with a brown recluse spider bite on the middle of his forehead (which made his brother call him 'the unicorn'). So the thought that we are routinely swallowing spiders in our sleep is incredibly freaky to both of us. You can imagine that someone may have first told someone like Rachel that they were swallowing spiders because they wanted to scare them, and the thought was so creepy that it stuck around. A 1954 book about insect folklore contained this myth, and most versions that are passed around imply that you are swallowing these spiders while you sleep. In 1993, a magazine article described this swallowing spiders idea as a myth, 'a ridiculous belief of the sort that someone would actually believe'. And, in classic myth-spinning form, now people quote this article as a source documenting that people actually do swallow spiders every year (even though this is the exact *opposite* of what the author was saying).

How do we know that the average person is not swallowing eight spiders a year? Can we prove that this does *not* happen? No. There are no studies to prove that this does not happen, but there are also no studies to prove that this *does* happen. We could not find any studies documenting any instances of people swallowing live spiders accidentally. And without any science or evidence, there is no reason to believe that this is actually happening. Furthermore, there are a lot of reasons why it is virtually impossible that people are swallowing so many spiders.

We'll be the first to admit that we are 'people' doctors and not experts on spiders. Therefore we must turn to 'spider' experts, such as the frighteningly dedicated aficionados at www. spiderzrule.com. (This site must be seen to be believed.) Why don't we routinely swallow spiders in our sleep? First of all, most people roll around in their sleep. This rolling around would probably scare the spiders from wandering anywhere close to your face. Second, you would need to have your mouth open, and not everyone keeps theirs open while they sleep. Third, experts tell us that spiders, like other arthropods, instinctively flee from open, breathing mouths. This makes sense – if you are an arthropod, unless you are suicidal, you are programmed to try to avoid things that might eat you. Finally, the spider would have to walk into your mouth and stay there in such a way that your swallowing reflex was triggered. We do not automatically swallow every time something goes into our mouths. So the chance that all of these things would happen together – that there would be a wandering, potentially suicidal spider in close vicinity to your mouth and that they would actually wander in to the wet, dark, breathing space and trigger your swallowing reflex – is really incredibly small.

Still, some will claim that a spider could fall into your mouth. What if it was hanging from the ceiling just above your mouth? The odds of this happening are also incredibly small. Your mouth is a relatively small target for a spider to hit randomly. And before you start worrying that eight spiders are going into your stomach every year through your nose, it's even more unlikely that you would swallow them this way than if they actively chose to crawl into your mouth. Your nostrils are (we hope) even smaller than your mouth, and thus an even smaller target for the spiders to hit. The most likely scenario if a spider actually went into your nose is that you would sneeze it out. The sneeze reflex is very sensitive and recognizes when even a small piece of dust gets in the nose.

It is possible that you have swallowed a spider, perhaps even more than one. But it is incredibly unlikely that you have swallowed eight in the last year. Or that enough people are swallowing spiders that each person would average eight (or any significant number) in a year. It's a myth all around.

Myth: You should move your bowels at least once a day

As paediatricians, we often hear from concerned parents that their child is not pooing every day. They worry that not having a bowel movement every day means that their child is constipated. Our colleagues who take care of older patients tell us that adults have the same concerns. What are you worried about? Do you secretly fear that your bowels might explode? Or that you will be bunged up for ever if you go too long without going to the loo? You are not alone. When you research bowel health on the web, you can find plenty of people who recommend the importance of moving your bowels at least once a day.

This is only a half-truth at best. While pooing regularly does prevent the discomfort and potential complications of constipation, you will be just fine if you don't move your bowels every day. A doctor will not define your problem as constipation until you have fewer than three stools a week.

How do we define constipation in adults? Get ready, because this may be more than you ever wanted to know about how much you need to defecate. Doctors say that you have chronic constipation if you have two or more of the following problems for at least twelve weeks of the year: you strain for at least 25 per cent of your bowel movements; you have lumpy or hard stools for at least 25 per cent of bowel movements; you don't feel like you have completely emptied out your bowels at least 25 per cent of the time when you defecate; you feel like your bowels are blocked up at least 25 per cent of the time when you defecate; you have to use your fingers to get the stool out at least 25 per cent of the time; and finally, you have fewer than three stools a week.

What about children, you ask? Surely doctors must want babies to poo more often. Actually, even small babies don't need to poo every day. Babies who poo only a few times a week could be just fine. How do we define constipation for a child? There are lots of medical groups that have come up with definitions. One of these definitions is known as the ROME 2 criteria for children, and they say that children with chronic constipation have at least two of these indicators: (1) at least two weeks of having a pebble-like, hard stool for the majority of their stools; (2) firm stools two times or fewer per week; and (3) no other bowel diseases. Another definition comes from the North American Society for Pediatric Gastroenterology, Hepatology, and Nutrition group: kids are constipated when they delay or have difficulty in defecation for two weeks or more, and this is significant enough to cause the child distress. A final definition is known as the Paris consensus, and it defines constipation as two of the following experiences for a period of eight weeks: defecating fewer than three times a week; losing bowel control because of constipation and leaking more than once a week; passing large stools that clog the toilet; having a stool you can feel in the belly by pressing on the stomach; intentionally withholding poo or not wanting to defecate; or experiencing pain when pooing.

If you or your child meets one of these definitions, you should, of course, talk to your doctor. But clearly none of the experts are worried if you don't move your bowels every day.

Myth: Your urine should be almost clear

When was the last time you examined the colour of your urine? In our experiences as doctors, we are often surprised by how much people can tell us about how their pee looks. If you listen to the 'experts', you probably think that your urine should be pale yellow or almost clear, and if your urine is darker than that, you should be concerned about being dehydrated. If you have yellow pee, you must need more water – think about the six to eight glasses of water a day. Right?

This is really just a half-truth. It is true that your urine gets darker when there is less of it. When your body needs to keep in more fluids (such as when you are dehydrated), you will pee less (a smaller volume of urine), and that urine will be darker. But having yellow urine does not mean that you are dehydrated! The colour of the urine depends on its osmolality, which is the technical term for how much stuff is dissolved in the liquid. When there is more stuff dissolved in a given volume of urine, the urine is more concentrated, and therefore it looks darker. However, the osmolality of normal urine can be very different from person to person. One specialist in fluid regulation tested the urine of sixty-nine healthy young adults and measured the average volume and concentration of their urine. For this group, all of whom were considered well-hydrated and healthy, the average volume of urine was 1,500 mL/24 hours and the average concentration was 600 mosmol/kg H_2O. At this average concentration, urine is yellow in colour, which could be interpreted as 'dark' compared to the 'clear' or 'pale yellow' you may think of as your goal. This concentration does, in fact, correlate well with a normal blood concentration that is far from dehydration. So, most of the time, normal urine from a healthy, well-hydrated person may be very yellow. If you are struggling to get it clear or pale yellow, you are probably worrying for nothing.

There is one caveat to dismissing this pee-colour business. Sometimes doctors will suggest that people with medical conditions such as repeated kidney stones make an extra effort to dilute their urine. If a doctor has recommended that you try to get your urine lighter than normal because of a medical condition, then you should listen to their health guidance on this particular issue. For those of you who are perfectly healthy, don't worry if your urine is yellow! If you are thirsty, drink. If you are not, don't worry about it!

Myth: You lose most of your body heat through your head

You probably heard this one from your mother. You were about to head out into the cold, and she called out that you had better remember your hat. After all, you lose most of your body heat through your head. Even the U.S. Army field manual for cold-weather survival says that you absolutely must cover your head in cold weather because 40 to 50 per cent of your body heat is lost through your head.

If this were true, we could walk around in the cold in just a hat and no trousers. But that would almost certainly leave you much, much colder than going without a hat. And much more exposed. And in much greater danger of a fine for public indecency.

This myth probably originated in a military study fifty years ago, when scientists put subjects in arctic-survival suits (without hats) and measured their heat loss in extremely cold temperatures. Well, since the only part of their bodies that were exposed to the cold were their heads, that's the part of the body from which they lost the most heat. Dr Daniel Sessler, Chair of the Department of Outcomes Research at Cleveland Clinic and an expert in hypothermia, says that if you repeated this study with the subjects donning only bathing suits they would not have lost more heat from their heads than from any other part of their bodies of proportional size. A more recent study from the US Army research environmental lab confirms that there is nothing special about the head and heat loss – any part of the body that is left uncovered loses heat. No matter what you leave uncovered, an exposed body part that is left out in cold weather will cause a drop in your core body temperature.

Myth: You can beat a breathalyser test

Some say that if you have had too much to drink and you get pulled over for a breathalyser test, you should suck on a copper coin. The theory is that the copper in the penny will create a chemical reaction with the alcohol in your saliva which results in an inaccurate breathalyser reading. Ironically, even the idea of a copper penny is largely a myth because most coins are made of other metals. But even the slight amount of copper within the penny cannot cause a reaction with the alcohol in your saliva. A breathalyser measures your blood alcohol content by examining the alcohol level of the air from deep within your lungs. The amount of alcohol in the air down there is actually very close to the amount of alcohol in your blood.

A penny seems to be the most commonly cited magic bullet for beating a breathalyser, but sometimes you will hear recommendations to suck on cough drops, peanuts, curry powder, onions, mouthwash or breath mints. The American TV show *Mythbusters* actually tested whether there were any things that you could put in your mouth or eat or suck on that would change your breathalyser test. They found that pennies, breath mints and onions did nothing to decrease the blood alcohol reading from the breathalyser, and using mouthwash actually increased the alcohol reading.

Back in the real world, one intoxicated man even thought that a mouthful of his own faeces would stump the breathalyser; his blood alcohol level was found to be twice the legal limit. We think that most people would have to be even more drunk than that to lean over, poo in their own hand and then stuff it in their mouth.

Often, the procedure for breathalyser tests is to do two of them, fifteen to twenty minutes apart. Any effects you miraculously got from what you had in your mouth the first time would need to be repeated for the second test, too. Even if you somehow managed to beat the system once, it's less likely that you would be able to do so again. Careful studies of many samples of breath alcohol samples show that they are incredibly accurate and correlate very well with blood alcohol levels – which are unaffected by what you put in your mouth. Even faeces.

Myth: You should never wake a sleepwalker

While Rachel sleeps in the same position all night, other people get a lot more exercise while sleeping. Sleepwalking is most common among fifteen- to twenty-four-year-olds and among those who also talk in their sleep. Having a psychiatric or mental illness also seems to increase your risk of being a sleepwalker. If you have a sleepwalker in your family, you may worry about them hurting themselves as they wander around the house or, even worse, outside. But many people try to never wake these sleepwalkers because they fear that it will cause a heart attack, shock or brain damage.

No sleepwalker has ever died as a result of being woken up while sleepwalking. Waking up a sleepwalker may confuse or frighten them, and they may even become violent in their confusion and try to hurt you. There have been rare cases of sleepwalkers committing murder while sleepwalking, so if you can guide the sleepwalker back to bed without waking them up, that might be the best strategy!

That does not mean that you should avoid waking the sleepwalker at all costs. If you do wake them up, they are not at any increased risk of heart attack or stroke or any of those other scary things. You may need to take action to prevent a sleepwalker from hurting himself, since they are not aware of their movements. If the sleepwalker is in danger and you cannot otherwise get them back to bed, you should certainly wake them up!

There are some steps you can take to prevent the sleepwalker from injury. Remove any dangerous objects from their room, put bolts on the doors and windows if you need to prevent them from opening them during sleep, and think about whether it might make sense to have the sleepwalker in a room on the ground floor so that they won't accidentally fall down the stairs.

PART II

'Do you want to catch pneumonia out there?'

Myths About Illnesses and Injuries and How We Treat Them

Myth: Cold or wet weather makes you ill

Many people believe that going out in the cold makes you more likely to get ill – including your mother.

One scientific paper reviewed all of the experiments where scientists tested whether we are any more likely to get ill when we are cold. In one of these studies, they actually put the virus that causes the common cold into people's noses and then chilled some of these people. Those who were chilled were no more likely to be infected with a cold than those who were not chilled. Another study looked at the relationship between chilling your feet and experiencing cold symptoms. The people who had their feet chilled were more likely to report cold symptoms afterwards. But does this mean that the cold feet caused the cold symptoms? No. It is very possible that these people just reported these symptoms because their feet felt so cold. There is no clear cause and effect.

There are other reasons why people may associate illness with cold weather. Some experts theorize that cold weather makes people more likely to stay indoors together, thus spreading colds and other viruses. There is no hard science to back that up, but we do know that colds are passed through close contact between people.

Myth: You can get a hernia by lifting something heavy

When you lift a heavy box, are you ever concerned about getting a hernia? Well, guess what? This is, at best, a half-truth. And it brings us back to a crucial point of distinction for interpreting science and understanding how things happen: there is a difference between causation and association. Causation makes something happen; association means that two things go together, but not necessarily that one thing caused the other to happen. Hernias and heavy boxes happen to be associated with one another, but lifting heavy boxes doesn't cause a hernia.

The other picky-doctor detail here has to do with the definition of a hernia. A hernia is technically an opening or a weakness in the muscle wall of the belly or groin. That opening or weakness is there whether or not you lift heavy things. This opening or weakness is not there because you lifted anything. Lifting did not cause the hernia. Now, here is where the association comes in. When you lift something heavy, it increases the pressure in your abdomen. If you already have a hernia, this type of increase in pressure can cause some of the contents of your belly (a little sac of your intestines, in fact) to bulge out through the hole. Lifting can cause this bulge to pop through or become more prominent, and so this makes your hernia more obvious to you. But even if you never lifted anything or engaged in any form of exertion, that defect or opening in your muscle wall would still be there. In other words, your hernia would still be there.

Exertion is clearly not the only, or even the most important, factor involved with hernias. In two studies investigating what people think caused their hernias, it is clear that the majority of people with hernias do not think that exertion or heavy lifting was involved. In a study of 129 patients with 145 hernias, only 7 per cent reported that they thought their hernia was attributed to a muscular strain or exertion like heavy lifting. In another study of 133 patients with 135 hernias, 89 per cent reported a gradual onset of symptoms for their hernia. Of the 11 per cent of people in the study who did think that the hernia was tied to exertion, the medical examiners could find no evidence to support that claim. Lifting something heavy and discovering you have a hernia may go together, but it was probably not the lifting that put that hole in your muscle wall.

Myth: You can catch poison ivy from someone who has it

Poison ivy creates a horrible, itchy rash that oozes across your skin and makes you miserable. If you have poison ivy, you'll notice others avoiding you like the plague. How contagious is poison ivy? Can you really catch it from someone else?

The oil (urushiol) from the poison ivy plant is, indeed, incredibly contagious. If that oil is still on your clothes or your skin or anything else, someone who touches it can get the rash too. Even if the oil is dried, it can still make your skin react. But once that oil is washed off, you are no longer contagious. No matter how bad your rash looks or spreads or oozes, the rash itself is not contagious. And it's normal for the rash to keep spreading even days after your contact with the plant. This is a delayed reaction and does not mean it is any more contagious – it's simply how your body reacts. It is normal for the rash to first appear twenty-four to forty-eight hours after contact with the plant oils. Scratching the blisters will not spread the rash. The fluid in the blisters is not contagious either. You cannot spread poison ivy from one part of your body to another through the rash.

However, even the dead plants can contain active oil for a long time and can give you a rash. Wearing clothing does not protect you because the oil can stick to clothes, which you may then come in contact with. The severity of the rash depends on the sensitivity of your skin and the concentration of the oil.

Leaves of three, let them be! Take precautions to avoid poison ivy plants in the first place if you really want to avoid this nasty rash.

Myth: Poinsettias are toxic

Poinsettias are widely used in festive decorations, but many fear that this 'toxic' plant could harm unsuspecting children and pets. Even though public health officials have reported that poinsettias are safe, many people still think poinsettias are poisonous. The largest study of poinsettia 'toxicity' to date involved an analysis of 849,575 plant exposures reported to the American Association of Poison Control Centers. None of the 22,793 poinsettia cases revealed significant poisoning. No one died from touching or eating poinsettias, and more than 96 per cent did not even require treatment in a healthcare facility. In 92 of the cases, children ate large amounts of poinsettias, but none needed medical therapy. In one study that looked at the effects of eating poinsettias on rats, the rats could eat massive amounts of poinsettia without being poisoned; they could even ingest doses that would be the same as a person eating 500–600 poinsettia leaves or 680 grams of poinsettia sap. The 'toxic' poinsettia is a myth. While you should always seek medical advice if someone eats a plant, you probably won't even need to see a doctor if you eat a poinsettia.

Myth: You should put butter on a burn

Burns are scary. When you or someone you love is burned, you want to take care of it as quickly as possible. Burns hurt like hell, and there doesn't seem to be much you can do. Many people believe that butter should be applied to the burn immediately, thinking it will help the burn to heal or prevent infection.

This is a myth, maybe even an outright lie. Butter could be one of the worst things to put on most burns. Butter holds in heat, which may make the burn worse. Furthermore, the butter could cause the burn to become infected since bacteria love to grow in something they can eat. Finally, the butter may make the burn hurt more, and it will make it more difficult for a doctor to examine the burn and see how bad it really is.

Don't use butter. Or lard, or milk or anything else you find in your fridge. There is only one case when you should even let the thought of butter cross your mind when someone is burned. If you are coated in hot tar or asphalt, you will feel better if you can get the tar off. Butter is something readily available that will work to remove burning tar. A combination of mineral oil and water works too. (You should also think about going to the doctor if you are coated in hot tar. In fact, an ambulance driver would probably be very pleased to come to your home and take you right to the hospital.) But unless you are coated in hot tar, do not put butter on your burn!

What should you do to care for a burn? There are several helpful options, none of which involve dairy products. First and foremost, seek medical attention. The correct treatment for a burn depends on how deep it is and how much area it covers. Your doctor can tell you the best thing to do. Ninety-five per cent of burns are minor and do not require hospitalization or complicated care. But there are some important steps to follow for superficial burns. First, take off any hot or burned clothing right away so it doesn't burn the skin any more. Cool the area, but *not* with ice. Ice is also a bad idea. It belongs in the fridge with the butter. Ice can cause further harm to the already damaged skin by freezing it. Use lukewarm or moderately cool water to cool the area. Clean a superficial burn with mild soap and tap water. Don't use disinfectants, because they can actually inhibit healing. You don't need to put any dressing on a superficial burn. You may need a tetanus jab if the burn goes at all deep. Topical creams are recommended to prevent infection for some burns, so consult with your doctor to find out whether any additional treatment is needed. Aloe vera may also help with healing and pain relief. And medicines to help with pain (like ibuprofen) are a fine idea, too.

No butter!

Myth: If you have allergies, you should own only a short-haired or non-shedding dog

We all know a child who begs and begs for a dog but cannot have one because someone in the house is allergic. Maybe you were that child. If allergies are a problem in your house of would-be dog lovers, you've probably heard lots of advice about getting a short-haired dog or a non-shedding dog. Allergic people often try to avoid long-haired dogs or heavy shedders, and some go so far as to keep their dog shaved in the hopes that allergies will be reduced.

The truth is, while you may be more allergic to one dog than another, the difference won't necessarily have anything to do with the hair. One study found that the breed of dog and the oiliness of the dog's skin both made a significant difference in how much dog allergen (called Can f 1) was produced by the dog, but that the length of the hair had nothing to do with it. Interestingly, the breed with the most significant amount of allergen production was the poodle, a breed typically thought to be less allergenic. The lowest allergen producer was actually the Labrador retriever. A review of the studies of dog breed and allergen production concluded that there are actually no breeds of dogs that are reliably less allergenic than others – short-haired or non-shedding. Even in patients already known to have dog allergies, studies could not document reliable breed-specific allergies.

What is it about dogs that make them allergenic? It really has to do with their dandruff and spit. A dog's skin and saliva contain proteins that cause allergies in some people. And so, since the problem is coming from the dog's skin and saliva, it is not the length of the shed hair that should be of concern to people with allergies.

Myth: A dog's mouth is cleaner than a human's

You've probably seen those people who give their dog wet, slobbering kisses right on the mouth — or maybe you are one of those people! Many people defend this behaviour with the explanation that a dog's mouth is cleaner than a human's.

Although few actually pause to try to back up this belief, some claim that a dog's mouth contains special substances that ward off infection. The truth of the matter is that all mammals have these substances. Pretty much all saliva contains enzymes that help to protect the body from infection. Dogs don't have any special advantage over people in this regard. And dogs don't have cleaner mouths.

This myth probably originated from the medical literature. Early studies showed that wounds caused by human bites were more likely to become infected than dog bites. This led many people to assume that the mouths of dogs had fewer bacteria than the mouths of people. This, however, was a leap of faith, not of science. You see, later research showed that this belief was mistaken for two reasons. First, human mouths contain 'human' bacteria for the most part, and dog mouths contain 'dog' bacteria. Humans are much more likely to become infected by human bacteria, and vice versa. If you count the actual number of bacteria, a human mouth does not contain more bacteria than a dog mouth. They just have *different* bacteria.

The second problem with those early studies is the manner in which people are bitten by others. It turns out that most bites sustained by people from other people heal without incident. Most human bites don't cause any more problems than dog bites. One sort of bite, however, has a much higher risk of infection, and this skews the numbers. Bites on the hand, sustained through a 'clenched fist' interaction, are significantly worse, and account for many of the human-bite infections. In other words, if you get a bite on your hand because you punched someone in the mouth, this is a bad bite likely to get infected. This is probably due to the mechanism and depth of wound sustained by this interaction, not because of the dirtiness of a human's mouth. Human-bite wounds anywhere else on the body are no more likely to become infected than dog bites. If you punched a dog in the mouth, you might have problems too, but those have not been studied.

It comes down to how you define clean. If you believe 'clean' refers to the types of bacteria and their ability to cause infection in deep wounds sustained in bar fights, then dogs might win. If, however, like most of us, you define 'clean' to mean having fewer bacteria in your mouth and *not* having recently used your tongue as toilet paper, human mouths win out in the cleanliness department every time.

Myth: If you get stung by a bee, don't squeeze out the stinger

This one is a half-truth! If you get stung by a bee, you must get the stinger out. The stinger can continue to pump out venom, and that will make your reaction to the bee sting worse. Getting the stinger out as fast as possible is an important part of treatment. But, how should you get it out? Some people advocate just squeezing, while others say you need to scrape the skin with something flat to remove the stinger. One study, using volunteers who allowed themselves to be stung by bees, tested the different methods of stinger removal and how bad the reaction was. The study found that the reaction was no worse if you squeezed out the stinger compared to scraping it off. What did make the reaction worse was if you left the stinger in longer. Do the fastest thing you can to get that stinger out! It's simply not true that you should use a particular method to remove it.

Myth: If you get stung by a jellyfish, you should get someone to urinate on the sting

Rachel was once stung by a jellyfish in the Indian Ocean, and her husband wanted to pee on her leg to help with the sting. Should she have let him? The science does not tell us. A study reviewing what we know about treating jellyfish stings concludes that we do not really know what to do. Experts all over the world disagree. One study does show that if you are stung by a Portuguese man-of-war, putting the sting in very hot water may reduce your pain but putting ice on it does not. Some people recommend using vinegar on jellyfish stings because it has been shown to stop tropical jellyfish stingers from firing in laboratory studies. However, vinegar does nothing to ease the pain from stingers that have already fired and it might even make the firing worse for some types of jellyfish. Other remedies, including urine, have not been studied. There is no reason to think that urine would help. In fact, if the urine dilutes the salt water around the stingers, it may cause them to fire more. With no evidence that urine helps and the possibility that it may hurt, our motto in this situation would be: 'When in doubt, don't let someone pee on you.'

Myth: Citronella candles effectively repel mosquitoes

You can accept this claim for citronella as a half-truth. Citronella is a natural mosquito repellent that does work to keep mosquitoes away. Citronella oils and lotions can keep mosquitoes away from you, but they do not work as well as DEET. The most common DEET-based repellents that you can buy will protect you from mosquitoes for five hours, whereas the citronella products should work for one to two hours. What about citronella candles? One study tested how citronella candles stack up against ordinary candles or no candles at all. Those within a few feet of the citronella candle had 42 per cent fewer bites than those who were not near a candle at all. Those near ordinary candles had 23 per cent fewer bites. Candles may protect you from mosquitoes to some extent, especially if you stay very close and use a citronella candle. But the most effective way to protect you and your children from mosquitoes is still to use a repellent that contains DEET.

Myth: Mosquitoes that buzz by your ear don't bite

As much as we hate mosquitoes, many of us do not know that much about them. There are a few facts about mosquitoes that make it very clear that buzzing and biting have nothing to do with each other. First of all, that annoying buzzing sound is produced by the flapping of the mosquito's wings. All mosquitoes have wings, and so all of them buzz. If a mosquito is by your ear, you will hear it buzz. Second, only female mosquitoes bite humans. They actually need your blood to produce the eggs that lead to baby mosquitoes. Male mosquitoes do not bite you. Therefore not all mosquitoes bite, but all mosquitoes do buzz. The two have nothing to do with each other. If you believe that the buzzing mosquito will not bite you, you believe an outright lie.

Myth: Green mucus indicates a sinus infection

As paediatricians, we hear very detailed descriptions of snot. Parents are always eager to tell us the precise shade of the stuff coming out of their child's nose. If it was yellow one day but green the next, we will hear about it. If it has been clear or white or the most lovely shade of emerald green, we will hear about it. Why do we hear so much about mucus? Many people believe that the colour of mucus matters − a lot. In particular, if the mucus has gone from clear to yellow or − the worst of the worst − green, then you must be getting a sinus infection, which means you need antibiotics straight away, right?

You can find experts who argue that the specific colour of your snot or phlegm is a big deal. Many doctors make decisions based on mucus colour. Researchers have looked at what doctors do if they hear that a patient had mucus of a particular colour. They found that when doctors hear that mucus is clear, only 1 per cent will diagnose a sinus infection and only 8 per cent will prescribe an antibiotic. For the very same patient, when doctors hear that the mucus was coloured, 38 per cent will diagnose a sinus infection and 59 per cent want to give the patient an antibiotic.

Are these doctors right? Do you need antibiotics if your snot is green? No! The importance of mucus colour is a myth that even doctors believe. Scientific studies show that mucus colour does *not* predict whether you need antibiotics. In a study of children with green nasal discharge, 142 kids were randomly assigned to take an antibiotic, a combination of a decongestant and antihistamine, a placebo or nothing at all. Researchers followed children in both laboratory tests and clinical exams, and they found absolutely no difference, except that the kids taking the decongestant/antihistamine had more negative side effects. None of the drugs helped the kids. The kids who were given antibiotics for their green mucus did not get better any faster than the kids who were given other drugs. Another comprehensive review of coloured nasal discharge concluded that there is no evidence existing to date that antibiotics shorten the duration of an illness when green snot is a symptom. However terrible the stuff coming out of your nose looks in all its fluorescent-green splendour, antibiotics are not going to help.

Myth: Using underarm antiperspirants causes breast cancer

Public awareness of breast cancer is a very worthy goal. However, all too often, emails or other communications to promote awareness, or warn about cancer risks, devolve into myths and propaganda.

One popular myth is that using underarm antiperspirant can cause breast cancer. The claim is that antiperspirants prevent the body from getting rid of toxins through sweat. The toxins then become stuck in the lymph nodes, where they might cause cancer by mutating a cell's DNA. Another version of this myth states that shaving the armpits makes the risk of using underarm antiperspirant even worse, because it creates small cuts in the skin through which toxins can enter.

There are no published scientific studies to support the idea that antiperspirants cause breast cancer. In fact, studies of thousands of people found no association between antiperspirant use and cancer. In addition, smaller case-controlled studies found no link whatsoever between using antiperspirants and getting breast cancer. One group of scientists studied 813 women aged between twenty and seventy-four who were diagnosed with breast cancer between 1992 and 1995. They matched these cases with 793 women who did not have breast cancer but were similar in age. The scientists then interviewed the women in both groups about using antiperspirants, using deodorants, shaving with a blade razor and applying any products under the arm within one hour of shaving. While it was clear that the use of deodorant and/or antiperspirant was very common among all the women, they did not find any differences between the groups. The risk of breast cancer was not increased if the woman used antiperspirant or deodorant, if they used either of these products and shaved their armpits, or if they put these products on soon after shaving

People believe a few other half-truths about breast cancer. Contrary to what some people think, breast cancer is not the most common killer of women; heart disease kills more women every year than any other disease. Furthermore, breast cancer is not even the most common cancer that kills women. More women die from lung cancer every year than from breast cancer. We think that it is absolutely crucial to be aware of breast cancer, to support those with the disease and to continue to look for causes of breast cancer that we can eliminate, but quitting smoking will go much further towards preventing cancer than abandoning your antiperspirant.

Myth: Flu jabs can cause the flu

As the season for influenza (commonly known as the flu) approaches, recommendations for flu jabs emerge. There are plenty of good reasons to try to avoid getting the flu – it makes you feel terrible, with aches, a cough and a fever. You may even end up in the hospital. Every year, more people die from influenza than from all other vaccine-preventable diseases combined. In the United Kingdom, 3,000 to 4,000 people die from influenza each year, and 250,000 to 500,000 people worldwide die from influenza annually.

The National Health Service currently recommends the influenza vaccine (the flu jab) every year for the following individuals: adults who are sixty-five and older, health and social-work professionals, adults and children with long-term medical conditions, and poultry workers.

Someone may have told you that getting the flu jab would give you the flu. This is an outright lie. The flu jab uses a dead virus to protect you from influenza. Dead viruses cannot make you ill. Dead viruses cannot be resurrected to cause infections. They are dead. We know someone out there will probably be arguing with us right now – maybe someone who thinks that the flu jab gave them the flu in the past, but it didn't happen.

If you've ever thought that your flu jab caused you to get the flu, you probably just experienced a bad side effect from the vaccine. The vaccine can cause soreness, redness and swelling where you get the shot. Some people also experience some low-grade fever and aches. This is not the flu. This is just the lousy part of getting a vaccination (though it might save your life). Second, you might have got ill right after the flu jab coincidentally. You may have been exposed to any other virus around that time or even to the influenza virus itself before you were injected. When you get a jab and get ill at the same time, it is natural to put two and two together and assume that one caused the other. But, once again, this is the difference between causation and association. Even if they happened at the same time, one event did not necessarily cause the other.

There are a few people who should not get flu jabs. If you have an allergy to chicken eggs, if you have had a serious allergic reaction to a previous flu jab, if you had the rare condition Guillain-Barre syndrome after a previous flu jab, or if you have a moderate or severe illness with a fever at the time that you want the jab, you should not be vaccinated.

What about the newer nasal-spray version of the flu vaccine? The nasal-spray influenza vaccine does not contain the dead virus; it uses a live, attenuated virus. While the virus in this vaccine is not dead, it is a special, genetically modified version of the virus specifically designed *not* to cause infection. It can never revert back to the original virus that can cause infection (the 'wild type'). This has never happened – not in scientific studies or in the millions of people who have had the influenza nasal-spray vaccine since 2003. Still, some people worry that the nasal-spray version of the influenza vaccine can come out of your nose and be transmitted to someone else. Shedding of the vaccine from the nose can occur, but the amount of vaccine virus that comes from your nose is incredibly small – much less than the amount needed to infect someone. And, in many studies, transmission of this attenuated vaccine virus has only been seen in one person. One child (in a study of 197 children) had influenza from someone else's vaccine detected in their nose in a single day, but it never caused any symptoms. In other studies, no one transmitted the vaccine virus at all. Even among HIV-infected children and adults, who would be at a higher risk for infection, no one was infected.

The moral of this myth–busting story: get a flu jab!

Myth: You should stay awake if you've had a concussion

Although they are usually mild and not a major cause for concern, concussions can be quite scary. The brain, which is made of soft tissue, is surrounded by fluid and contained within a hard case – your skull. When you are injured, especially forcefully, it's possible for your brain to smash into the wall of your skull, leading to bruising or potentially even bleeding. When this happens, your brain can stop functioning for a period of time. That is a concussion.

Concussions are fairly common (Rachel has had one). The World Heath Organization Collaborating Centre Task Force on Mild Traumatic Brain Injury estimated that in the general population concussions occur in more than 6 out of every 1,000 people. However, most concussions are mild and require no treatment.

Many people believe, and have been told again and again, that it is important to keep someone who has had a concussion awake. There is a mistaken belief that someone who falls asleep after sustaining a concussion might fall into a coma and not wake up again. The origins of this myth are not well understood, but are probably to do with what's called the 'lucid interval' experienced after some very severe head injuries. After a bad head injury, there can be some slow bleeding. Once the initial shock of the injury has worn off, a patient can appear normal for a period of time (the lucid interval) while the bleeding continues. Once enough bleeding has occurred, the brain is compressed, and emergency surgery may become necessary.

However, as scary as this sounds, bleeding occurs in a very, very small number of head injuries. Mild concussions usually resolve quickly, without any effects whatsoever. In fact, doctors believe that athletes can return to a game within fifteen to twenty minutes after a very mild concussion. There is no reason to believe that bleeding occurs as a result of these injuries, and therefore no need to watch concussed people closely. Those with more worrying symptoms, such as worsening headaches, nausea, confusion, difficulty walking or talking, vomiting or even seizures, should seek immediate medical attention. Any of these symptoms mean you should go to hospital for evaluation.

The bottom line is that if you have a concussion that is deemed mild enough to warrant no medical attention or no treatment after evaluation, then the risk of bleeding is probably negligible and there is no reason why you shouldn't go to sleep.

Myth: If you agree to donate your organs, doctors won't work as hard to save your life

Many people are in desperate need of an organ transplant to save their lives or to improve their health. In fact, almost 7,700 people are on the UK organ-transplant waiting list right now. Although an average of nine people get an organ transplant every day, many more never get the phone call to say the organ they need has been found. Thousands of people die while still waiting.

Most of us don't like to think about what would happen to us or to our loved ones if we died or were in a near-death situation. Organ donation does cause some people unnecessary fears, like the idea that if you agree to donate your organs doctors will not work as hard to save your life. This is a common concern about organ transplantation, and one that may be particularly strong in certain ethnic groups, but it's simply not valid.

First of all, when you go to the hospital for a life-saving treatment, your doctor's highest priority is to keep you alive. You will be seen by the doctor who has the best skills to treat your medical problem (and this is unlikely to be the organ-transplant doctor). Second, doctors usually have no idea whether you have agreed to be an organ donor until you have actually died! Neither of us has ever witnessed a doctor ask about organ donation until there was absolutely nothing left to do clinically. The doctors wait until you are dead to decide about organ donation. In fact, if you have agreed to be an organ donor, you will probably have even more tests done to determine that you are officially dead than someone who is not donating an organ.

Perhaps the most convincing argument that being an organ donor will not affect your medical care is that medical doctors are more likely than the general public to make themselves organ donors. In fact, studies suggest that the more education a doctor has had, the more likely he or she is to be an organ donor. Healthcare professionals who know a great deal about organ donation are more likely to be organ donors themselves and to recommend organ donation to their relatives. Now, if the doctors who know the most about what happens during organ transplantation are also the most likely people to sign up to do it themselves, don't you think that they are reasonably confident that being listed as an organ donor will not hurt the quality of their medical care? Doctors do not always get things right, but if they have made this choice for themselves, then it's a good bet that you can feel comfortable making this choice as well. More importantly, you might even be able to save the life of one of the 7,700 people on that long waiting list.

Myth: If you use the highest SPF sun cream, you won't get burned

Let's get one thing out of the way: you absolutely should use sun cream. Exposure to the sun's ultraviolet rays (UVA and UVB) causes damage to the skin which can lead to skin cancer. UVB rays are the ones that cause most sunburn, but UVA rays, which penetrate deeper into the skin, also cause the skin to wrinkle, sag, get leathery and all those other signs of aging we would like to avoid. Sun creams with an SPF of 15 or higher really do protect your skin against the UVB rays, and most protect against UVA rays too − so use your sun cream!

If you really want to protect your skin, is the key to layer on the highest SPF you can get your hands on? Not necessarily. First of all, there is a 1.6 per cent or less difference between sun creams with SPF 30, 45, 50 and 60. Sun creams with SPF 15 block approximately 94 per cent of all incoming UVB rays. Sun creams with SPF 30 block 96 per cent of the UVB rays. Sun creams with SPF 40 block 97 per cent of the rays. The higher SPFs do block more UV rays, but it is not clear whether they are increasingly effective (in actual use) over the SPF 50 mark.

If you rely on the SPF model, you can still get your skin into lots of trouble. No sun cream, regardless of its strength, is effective for longer than two hours without reapplication. You'd be better off faithfully reapplying your SPF 15 or 30 sun cream every two hours than staying out in the sun all day with SPF 50-plus. Plus you need to use *a lot* of sun cream for it to be effective. You should use at least one ounce, the equivalent of one full shot glass, on the entire body. Sun cream SPF is tested using 2 mg of sun cream per square centimetre of your body, which equates to two fingers-length of product applied to each of the eleven areas of the body (i.e., head, neck and face; left arm; right arm; upper back; lower back; upper front torso; lower front torso; left upper leg; right upper leg; left lower leg and foot; right lower leg and foot). If you stayed at the beach all day, you'd have to use up almost an entire 200ml bottle of sun cream in order to follow this recommendation. Since most people don't even come close to following the official recommendation, the level of protection they receive from their sun cream is probably half, or less than half, of what is described on the bottle.

There are other rules for sun cream application that you should follow to make sure you're as protected as possible. You should apply your sun cream thirty minutes before you go out in the sun to let the ingredients bind to the skin. Ideally, you should apply it again twenty minutes after you go out. A study shows that this early reapplication is even better than reapplying two hours later (which is the standard recommendation). If you follow this guideline, you need to reapply the sun cream again only if you go swimming, dry off with a towel or sweat excessively.

Simply put, once you get to SPF 15 or 30 there is not a huge difference between the levels of sunscreen. There is no harm in using the highest possible SPF, but what will really make the difference for your skin is applying a thick layer of sun cream thirty minutes before you go outside and then again twenty minutes after you've been outdoors.

Myth: Vitamin C, echinacea and zinc will keep you from getting a cold

Everyone hates being ill. And when our offices and clinics are packed with sneezing, sniffling, coughing families, we desperately wish there was something effective we could recommend that would prevent the colds and coughs. Some may tell you that taking vitamin C or echinacea or zinc will keep you from getting a cold. Unfortunately, science does not suggest that any of these things are going to help you very much.

Vitamin C:

Vitamin C was discovered back in the 1930s, and people have been suggesting that it helps alleviate respiratory illnesses and colds ever since – and we can't blame you for believing that it works. After all, two-time Nobel Prize-winner Linus Pauling thought the same thing and told everyone to load up on vitamin C. Are we claiming to be smarter than a Nobel Prize-winner? No. (Well, maybe Aaron is.) What Dr Pauling did in the 1970s was exactly what we are doing now – he looked at the scientific studies of vitamin C and saw that the results indicated that vitamin C worked to keep people from getting colds. The advantage that we have over Dr Pauling is that many more studies have been done since that time. When the results are combined, the science tells us much more definitively that taking vitamin C every day will *not* prevent a healthy person from getting a cold.

However, if you run marathons every weekend, routinely plummet down Alpine slopes or engage in sub-arctic military training exercises, then be our guest and take vitamin C regularly.

Only those individuals exposed to short periods of very extreme physical stress, extremely cold conditions, or both, seem to have less of a chance of getting colds if they take vitamin C regularly.

What about all the people out there who aren't super-human athletes training in extreme temperatures? If you take vitamin C every day, your cold might last a little shorter (we're talking *hours* shorter – not days). But if you start taking vitamin C *after* you feel ill, your cold will not be any shorter or any less severe (based on the results of eleven studies of more than 6,000 people). Most studies that examine vitamin C efficacy are randomized placebo-controlled trials. In a review that combined thirty of these great studies and included more than 11,000 people, taking 200mg of vitamin C or more per day was shown to be completely ineffective in preventing illness. For the average Joe, there's no need to bother with the extra vitamin C.

Echinacea:

Echinacea has been so widely touted as a cold treatment that it has become one of the most commonly used herbal products. The science and studies of echinacea leave us with some questions. What is clear to us is that echinacea does *not* cure colds. Colds are caused by respiratory viruses, and there is no science to suggest that echinacea gets rid of these viruses. However, it is possible that echinacea might shorten the lifespan of your cold. The conclusion of a 2006 Cochrane Systematic Review was that echinacea does not work. When the Cochrane Systematic Review looked at sixteen studies of the herb, echinacea was no better than a sugar pill at preventing colds. One of the studies was published in the prestigious *New England Journal of Medicine*; doctors examined almost 400 volunteers taking either echinacea or a sugar pill and found that the echinacea did not change the severity of your cold symptoms or how the infection progressed.

Some of these studies did suggest that echinacea works to shorten colds in adults when taken at the very first sign of a cold. In children, echinacea was not found to prevent or shorten colds.

The research didn't stop there. After the Cochrane Systematic Review was published, two other groups of researchers tried to compile newer studies to answer the question. Because many of the studies used different techniques, it is difficult to compare the results from one study with another. One group looked at studies where researchers actually put gobs of the cold virus right into the study participants' nostrils and then checked to see if the herb did anything to stop the individuals from getting ill (doesn't that sound like fun?). Echinacea didn't work in those studies.

In 2007, yet another group of researchers examined existing studies, and they concluded that echinacea works. They looked at fourteen studies, and decided that echinacea reduces your chance of developing the common cold and shortens the length of time you experience cold symptoms by almost a day and a half. Unfortunately, other researchers have criticized this study and don't agree with its conclusions. The researchers combined studies that might be too different from each other to put together – they tried to compare apples and oranges. We are not necessarily saying this study is wrong, but we do think that it might not give us the best answers and so we are not yet convinced that echinacea really works. Echinacea is not the cure for the common cold, but it may help you feel a little better.

Zinc:

Once you've given up on the vitamin C and the echinacea, you might decide to try zinc. Unfortunately, there is no evidence that zinc will cure your cold.

The results of zinc-efficacy studies have been very inconsistent. Some studies have found that zinc lozenges are helpful and some say just the opposite. In a systematic review that examined fourteen of the best zinc studies, half of those studies said that the zinc did nothing at all and half reported that the zinc helped. How do we decide which to believe? The quality of the studies was evaluated using a set of eleven strict criteria. Only four of the studies met all eleven of these criteria and could be called well-designed. Looking only at the four well-designed studies, two reported no effect from zinc lozenges, one reported no effect from zinc nasal spray, and only one reported a positive effect. Three of the four good studies did not find zinc to be effective.

The one good study that did show a positive effect for zinc was a study of zinc nasal gel, which was found to improve your cold symptoms and reduce how long they last. The only problem – and it's a big one – is that putting this zinc nasal gel in your nose ultimately resulted in a $12 million settlement in 2006 from 340 lawsuits claiming that the zinc nasal gel damages the sense of smell for users. Unless you are willing to give up your sense of smell (and your sense of taste, which goes with it), don't put zinc in your nose!

You also might be interested to know that edible zinc (zinc lozenges) often don't taste very good. Plus the most common side effect of zinc lozenges is nausea. Even if you can bear the wretched taste and the sick-to-your-stomach feeling, zinc has only a small chance of helping you feel better. And the two best studies of zinc lozenges showed that it won't help your cold at all. We suggest you ditch the zinc.

Myth: Breast milk can cure ear infections

Not only is breast milk uniquely formulated to meet the nutritional needs of infants and young children as they grow but it also contains a mother's antibodies against infection and may help to decrease a baby's chance of contracting certain types of infection. Some advocates of breastfeeding suggest that breast milk can be used to ease all sorts of infections and problems that infants deal with, and they even go so far as to recommend that mothers put a bit of breast milk in their child's ear to ease ear infections and ear pain. They also recommend breast milk in the eye for eye infections or pink eye. Further uses suggested for any spare breast milk you might have around the house include clearing a stuffed-up nose, soothing nappy rash, easing a sore throat, removing make-up and healing mosquito bites.

As paediatricians, we absolutely endorse breastfeeding. Good scientific studies have linked breastfeeding to lower rates of ear infections, infections that cause diarrhoea and vomiting, eczema, asthma and obesity. Although these studies cannot prove that breastfeeding prevents these conditions (it's unethical to do a randomized controlled trial of breastfeeding), the studies do show that breastfeeding is significantly associated with a lower risk of disease among babies and mothers. Breast milk contains both the white blood cells that gobble up germs (macrophages) and a protective protein (immunoglobulin A) that coats the lining of a baby's stomach and can help prevent germs getting through. Breast milk also contains antibodies that the mother's body has made against particular germs. Also, there are several ways that breastfeeding (putting the breast milk in the baby's *mouth*, not the baby's ear) can help protect against ear infections. The germ-fighting properties of breast milk might prevent infections in the baby in general. Breastfed babies are often fed in a more upright position than bottle-fed babies, and this can prevent milk from backing up through the little tube that connects the back of the baby's mouth to the ear (the Eustachian tube). Breastfed babies also have fewer allergies than bottle-fed babies. Allergies can cause a build-up of fluid in the middle ear, which can then get infected with bacteria, leading to an ear infection. If breastfed babies have fewer allergies, they might be less likely to get this fluid build-up and then less likely to get ear infections from that fluid.

However, despite all of the great benefits of breast milk, it is not meant to be put in a baby's ear. We could find no scientific evidence at all that recommended putting breast milk in an ear or an eye. Furthermore, the anatomy of the ear makes it highly unlikely that breast milk could help fight an infection. The ear canal is a tube with a wall sectioning off the outer ear canal from the middle ear canal. This wall is called the tympanic membrane (the ear drum). An ear infection occurs when fluid builds up behind that wall and gets infected. If you were to put breast milk into the ear from the outside, it could not enter the area of infection unless there is a hole in the ear drum (and there should not be a hole). Furthermore, putting breast milk into the ear canal tube could increase the chance that your child gets an infection in the outer part of his ear canal. Even though breast milk has some ability to fight germs, any sugary substance (which breast milk is!) could also provide a great place for bacteria to grow. (This is why we recommend that parents don't just leave breast milk sitting out on the worktop for ever. It needs to be carefully refrigerated or frozen so that bacteria are unable to grow.) The last thing you want when your child has a middle-ear infection is for them to also get an outer-ear infection from the breast milk you put in there.

Breast milk is great – when used as intended. So, if you want to decrease your baby's chance of getting ear infections, don't squirt it in his ear!

Myth: Acupuncture doesn't really work

We have talked about a lot of home remedies and alternative therapies that just don't seem to work when they are tested scientifically. But this is not to say that traditional herbal or complementary therapies never work. On the contrary, we are happy to recommend them if the science shows that they are effective. We have the same requirement for the usual therapies of 'Western' medicine. If the science says that something works and that it is safe, we recommend it, no matter where it comes from.

Acupuncture is an example where the sceptics from the world of doctors and Western medicine may be wrong, and the proponents of traditional or complementary therapies may be right – at least for some problems. Evidence suggests that acupuncture can be an effective therapy for treating nausea and treating pain in some patients. Researchers have subjected acupuncture to careful (and some not so careful) studies to see whether or not it works for a variety of medical problems. When researchers put all of the studies together and analyse them carefully, the science tells us that acupuncture works for some things and seems not to work for others.

If you are feeling really nauseous because of something like having anaesthesia or chemotherapy, acupuncture may be able to help you. Compiled studies looking at a particular pressure point in the wrist showed that acupuncture reduced nausea after surgery and anaesthesia even more than anti-emetics, the drugs intended to help you feel less nauseous. Acupuncture can also decrease the immediate vomiting some people experience after chemotherapy.

There is some evidence to suggest that acupuncture also helps with particular types of pain. Three trials of the use of acupuncture during labour suggest that it may decrease a woman's need for chemical pain relief when delivering a baby, although the results were only marginally significant. Randomized controlled trials showed that there may be some short-term benefits from acupuncture in relieving shoulder pain, and acupuncture worked better to relieve immediate neck pain than sham or fake treatments. The subjects who got acupuncture also reported less neck pain when they returned for follow-up a week later. Acupuncture seems to have some benefit for chronic back pain (but not for acute back pain), but most people agree that more and better studies are really needed. A number of these studies are now underway. Some studies also suggest that acupuncture works for headaches, but again, more studies need to be done before we can make a clear call on acupuncture for this problem as well.

Acupuncture is not a cure-all though. For some problems, equally careful studies and compilations of the literature tell us that it doesn't work or probably doesn't work. Acupuncture does not seem to work to treat cocaine addiction, depression, insomnia or irritable bowel syndrome. It may or may not help for tennis elbow; you may have some short-term pain relief if you get needle acupuncture for tennis elbow, but it doesn't seem to last long and the limited benefit is based on only two small studies.

PART III

'But I was on the pill!'

Myths About Sex and Pregnancy

Myth: Men think about sex every seven seconds

It is not really clear who first coined this phrase, nor why it focuses on men and not women, but it has become an accepted truth that men think about sex a *lot*. Depending on which version you've heard it could be as little as every few minutes or as much as every seven seconds. This fact is stated often, even by reputable media sources, in stories that describe the differences between men and women, claiming that men are, well, obsessed with sex.

On the face of it, this myth is simply ludicrous. Just think about it. If we assume that the average male is awake for sixteen hours a day, each man would have to think about sex over 57,000 times a day. That's about as many times as a person breathes while awake. Someone thinking about sex that often would be incapable of performing any other tasks at all. Moreover, thinking about anything that often would probably drive you insane.

Many attribute this 'fact' to the Kinsey Institute, which was founded right in our own backyard at Indiana University in 1947 and published its first important work, *Sexual Behavior in the Human Male*, in 1948. This has probably added to the myth's perceived validity, as the Institute is widely accepted as the final word on research related to human sexuality. But no data released by the Institute have ever shown that any human beings think about sex this often.

The most comprehensive survey ever completed on sexuality in America was published in 1994, titled *The Social Organization of Sexuality: Sexual Practices in the United States*. In response to the question: 'how often do you think about sex?', 54 per cent of men reported thinking about it every day or several times a day, 43 per cent a few times a month or a few times a week, and 4 per cent less than once a month. So although half of men do think about sex as much as several times a day, this does not even come close to the several times a minute of the myth. However, many men think of sex even less frequently – not even every day. And, of course, if you're interested in how women compare, here are the data: when asked the same question: 19 per cent of women reported thinking about sex every day or several times a day, 67 per cent a few times a month or a few times a week, and 14 per cent less than once a month. Women do think about sex less frequently than men, but the difference is not as great as most people believe. While men might think about sex every day, women are more likely to think about it a few times a week.

As always, it's important to remember that these are averages. Some men will think about sex more than just a few times a day and some a lot less. But none can possibly have the time, focus and mental stamina to think about it every seven seconds.

Myth: Condoms protect you from all STDs

We love condoms. If you use a condom, you decrease your chance of getting a sexually transmitted disease. However, many people think condoms are 100 per cent effective and that they simply cannot get a sexually transmitted disease while wearing a condom. Unfortunately, that's just not true.

Now, this should never be a reason not to use a condom. They are *very* effective in preventing the transmission of certain diseases. For instance, a review of studies done on condom use to prevent gonorrhoea found that using condoms reduced the risk of getting gonorrhoea somewhere between 30 per cent and 100 per cent for males and between 13 per cent and 100 per cent for females. This is great! Similar studies looking at protection against chlamydia found that condom use reduced the risk of getting chlamydia somewhere between 15 per cent and 100 per cent for males and 10 per cent and 100 per cent for females. Again, hooray for condoms! Perhaps the most concerning sexually transmitted disease is HIV. There is no doubt that the use of condoms also reduces your chance of contracting HIV, the virus that causes AIDS. Much of Rachel's research work involves HIV, and preventing HIV is the best reason she can think of to use a condom.

Condoms are great, but in many cases they are not perfect. A 2005 study examined how well condoms could prevent the transmission of herpes simplex virus. Of the 1,843 participants in the study, just over 6 per cent became infected. Those using condoms were less likely to become infected with one type of herpes, but condoms provided no protection against becoming infected with a different type of herpes. And, if the herpes is in a spot that is not covered by a condom, the condom helps you even less.

And there are other sexually transmitted diseases that might be harboured in areas not covered by condoms: such as human papillomavirus (which causes genital warts), crabs (which scurry all around), and genital ulcers (if these lesions are not covered by the condom, then ulcers can definitely be transmitted.

In the lab, condoms are almost 100 per cent effective in blocking the nasty bugs that cause sexually transmitted diseases. However, real life is not a lab. Condoms don't cover the entire genital area. Moreover, condoms can break, and if they do, you can pass on a disease. Plus we have yet to meet the mythical person who never, ever fails to use a condom – and unless you use a condom every single time you have sex, you're not fully protected from disease. Of course, if you are really and truly only going to have sex with one person (and only one person) in a completely monogamous relationship, and are disease free, you may not need a condom. In summary, condoms are not perfect, but there are many good reasons to use them and we recommend them despite their flaws.

Myth: Semen is loaded with calories

Even though we show in another chapter that more women engage in oral sex than you might think, some do not. One of the reasons often cited by those with an aversion to oral sex is actually a nutritional one. Many people believe that semen is high in calories. While this may seem crazy (or even disgusting to some), a rudimentary search of the Internet shows that it is an active concern for a surprising number of women, especially teenage girls.

We should start by making sure that everyone understands exactly what we are talking about. Semen is composed of sperm and fluid from the seminal vesicles, prostate gland and Cowper's gland. Sperm develop in the testicles and then move to the seminal vesicles through tubes known as the vas deferens. While in the seminal vesicles, sperm mix with the other fluids to form semen.

To figure out the caloric content, you have to know what to count. A normal male ejaculation is about 3 mls of semen, or less than one teaspoon. The amount of semen in any given ejaculation can be higher or lower, however, depending on how recently the last ejaculation occurred. In fact, one study found that each day of not having sex could increase the volume of semen by over half a millilitre. A normal volume of semen can contain about 75 million sperm per millilitre, so the usual ejaculation will contain somewhere between 200 and 300 million sperm.

But sperm are not where the calories come from. It's the fluid that has everyone concerned. Although there is little nutritional value to semen, it does contain water, vitamin C, sodium, potassium, calcium, magnesium, zinc, citrate, chloride, protein and (most concerning to those watching calories) fructose – a type of sugar. However, the amounts contained in semen are very, very small. In total, a typical male ejaculation probably contains about five to seven calories. This amount is so small that you would probably burn it off during any sexual activity at all.

This nutritional information should not be used in any debate about spitting versus swallowing. Just know that swallowing won't ruin your diet.

Myth: Single people have much better sex lives than married people

Married people bemoan the many responsibilities that put a drain on their sex lives. They look at their single friends and are secretly jealous of the more frequent and better sex they must be having.

However, that's not the case. *The Social Organization of Sexuality: Sexual Practices in the United States*, which is pretty much the final word on sexual practices in the US, tells us that, while 23 per cent of single men report not having had sex in the last year, only 1 per cent of married men say the same. Another 25 per cent of single men have only had sex a few times in the past year, compared to only 13 per cent of married men. In fact, a full 43 per cent of married men report having sex at least two to three times a week, compared to only 26 per cent of single men. You see, married men really are having more sex all around. In the UK, *The National Survey of Sexual Preference and Lifestyle* showed that married people had sex an average of seven times in the previous four weeks, compared to three times in the previous four weeks for everyone.

The numbers for women are similar. In the US, 55 per cent of single women have had sex a few times or less in the past year; only 15 per cent of married women say the same. On the other hand, 39 per cent of married women report having sex at least two to three times a week, compared to only 20 per cent of single women.

Another study, published by the National Opinion Research Center (US) in 2006, looked at the average number of times people had sex in a year. Married people had sex an average of 66.3 times per year, compared to 61.9 times per year for those who were never married. Even amongst those between eighteen and twenty-nine years old, who may be considered the most likely to have sex, those who were married had sex an average of 109.1 times a year, compared to 73.4 times a year in those never married.

Additionally, 75 per cent of married women reported 'usually' or 'always' achieving orgasm, while only 62 per cent of their single pals said the same. And the benefits of marriage don't end there – married people are also more likely to give and receive oral sex. While 71 per cent of men reported performing oral sex before marriage, that figure is around 80 per cent for married men. Similarly, only 62 per cent of women reported performing oral sex before marriage and 71 per cent reported performing it while married. Greater percentages of married women and men also report having received oral sex than single women and men.

Married people shouldn't necessarily envy the sex lives of their single friends – perhaps their single friends should be envying them!

Myth: Women lose interest in sex after menopause

Even among those who accept that women in general have more than a passing interest in sex, many believe that women's interest in sex goes way down after menopause. And many people assume that you do a lot less of it as you age.

The truth, however, is not so clear. First of all, good data exist on the amount of sex that people have at various ages. Once again we can turn to *The Social Organization of Sexuality: Sexual Practices in the United States*. About 75 per cent of women in their thirties have sex at least a few times a month. Women in their forties have less sex, but 69 per cent have sex at least a few times a month. And, when women hit their fifties, 48 per cent are still having sex a few times a month or more. Only 30 per cent of women in their fifties report not having sex in the last year. Lest we think that this is entirely an issue for women, however, 67 per cent of men in their fifties also report having sex a few times a month or more.

Yes, older women seem to have less sex than younger women. However, science suggests this is not because of the hormones that go with menopause. In a 2004 comprehensive review article in the journal *Menopause: The Journal of the North American Menopause Society* (they have a journal for everything, don't they?), researchers discuss how many different things must be taken into account when a woman has low sexual desire. The factors involved can include a woman's age, quality and length of relationship(s), physical health (including menopause status), sexual experience and mental health/personality. Some studies have found that the effect of menopause on sexuality is indirect; in other words, menopause symptoms can affect well-being, which affects how one responds sexually, which affects how often one has sex.

Other studies have found that the sexuality of older women depends more on other factors such as physical and mental health than it does for younger women. How healthy the woman is, not whether she has gone through menopause, plays a bigger part in how much pleasure and interest she has in sex. Often the lack of sex in a woman's life is due more to emotional problems, communication problems, or feelings of pain or guilt. And, of course, the older a woman is, the more likely that her partner has become ill or died, making sex less possible or pleasurable.

In summary, although older women generally have less sex than younger women, many are still doing it, and still doing it fairly often. Menopause is not necessarily to blame for the decreasing numbers who are doing it; older women may be less sexually active due to changes in their physical health, mental health or relationships.

Myth: Virgins don't have openings in their hymens

A lot of people don't know what the hymen really is – even doctors. People generally think that the hymen is 'something' in the vagina that seals it off until you have had sex or until you are no longer a virgin. This is not true.

First of all, the hymen is not technically in the vagina. It is a thin layer of tissue outside the vagina (in the vulva) that partially conceals or covers up the vagina. When a female foetus is developing in the womb, there is originally no opening or hole in the hymen, but an opening develops by the time the baby is born. Long before you lose your virginity, your hymen already has an opening.

There are some women who do not have an opening in their hymen because normal development did not occur in the womb. This is a problematic medical condition known as imperforate hymen. Once a girl begins menstruating, if there is no opening in her hymen, then the menstrual blood builds up in the uterus and vagina. This is not a common problem; it only occurs in 1 in 2,000 females. An imperforate hymen is rarely diagnosed until girls begin menstruating and they discover a big, painful mass in their belly around the time of their period (a mass of blood that cannot exit the vagina). In order to fix an imperforate hymen, a doctor must perform surgery to create an opening in the hymen.

Myth: A doctor can tell whether you are a virgin or not

Many cultures around the world value virginity, especially in girls. And because of the high value placed on virginity, people assume that there must be some way to tell whether or not you are a virgin.

First of all, many people assume that you can tell whether the person you have sex with is a virgin just because of how it may feel or whether or not there is bleeding. This just isn't true. But even if the person you are having sex with can't tell, most people believe that a doctor can tell, especially if she does a careful exam of the genitalia. This isn't true either.

Women's hymens are different shapes and sizes. And, as we explained in the previous chapter, the hymen should have an opening, whether or not the woman is a virgin. Even when doctors use incredibly careful examination techniques, they cannot necessarily tell whether someone has had sex. Half the time, women who have had sex have absolutely no changes detectable by a doctor using a special camera with 10x magnification and measuring the actual width of the hymen's opening to within a millimetre. In a careful study from 2004 of females aged between thirteen and nineteen, 48 per cent of the girls who admitted to having had sex had some small changes on the edge of the hymen. But 52 per cent of the girls who *admitted* that they *had* had sex showed absolutely no findings in even this very detailed exam. So, more than half of the time, when a girl has had sex there will be no changes. Therefore you cannot prove that someone is not a virgin based on what you can see.

And if doctors doing this kind of exam cannot tell if someone is a virgin, there is no way that you can tell either.

Myth: You can't get pregnant using the 'pull out' method

Close contact between a penis and a vagina can lead to pregnancy. Full stop. Even if you use the withdrawal method, it still may be too late.

Before a man actually ejaculates or climaxes, there are usually drops of semen at the end of the penis. These drops of semen help to lubricate the head of the penis and may be present before a man feels close to ejaculating. Even one drop of semen can contain a million sperm. And it only takes one sperm to get pregnant. There is a smaller chance a sperm will fertilize an egg when you start with 'just' one million sperm, compared to the chances with hundreds of millions of sperm, as in full ejaculation. But it's still possible for one of the sperm out of that drop of semen to make it to the egg. Furthermore, the seconds before climax are not the best time to expect someone to use good judgement and pull out. American studies show that when 100 women use this method to prevent pregnancy, 27 end up getting pregnant within a year. Even if you pull out perfectly every time, 16 out of 100 women get pregnant. French men and women did a little better with the 'pull out' method; in one study, 10 in 100 women using this method became pregnant. These odds still suck.

Other studies confirm that lots of people get pregnant using the withdrawal method. In a study of 1,910 women in Turkey (where 35 per cent of the group was using coitus interruptus or pulling out to try to prevent pregnancy), 38 per cent of the women experienced at least one unwanted pregnancy. In a study from a family-planning association, where about 30 per cent of the population was reported to use coitus interruptus as their method of choice, 34 per cent of people indicated that they or their partner had become pregnant when relying on withdrawal.

A woman can also get pregnant even if the man doesn't put his penis all the way into her vagina. If he puts it part way in and then ejaculates, a woman will have lots and lots of sperm inside her that can still try their luck at swimming up to an egg. Even if he doesn't ejaculate, the woman may have some small drops of semen in her vagina, with their millions of sperm, also swimming along and trying to fertilize the egg. Even if the drops of semen are right outside the vagina, there is a chance that a sperm will sneak inside and make its way to the egg. It only takes one!

Myth: You can't get pregnant during your period

For most women, the chance of getting pregnant during your period is much less than at other times of the month. But it is never, ever impossible. You will hear this refrain a lot in this section – it is never, ever impossible. Getting pregnant requires a sperm and an egg, so if those two things are around, you can have a pregnancy.

If you have short menstrual cycles or irregular periods, there is even more of a chance that your body might send out an egg (ovulate) or that there might be an egg around when your period starts. The egg can live for several days, and not all women ovulate weeks before their period. Some ovulate much closer to the time that the period occurs. Plus sperm can live for days, even up to a week, in the nice, wet environment inside a woman's body. Therefore the egg and the sperm could be around at the same time, which could lead to a pregnancy.

Some people attempt to use the 'rhythm method' to prevent pregnancy. In other words, they only have sex during the 'safe' days of the menstrual cycle, when the woman is less likely to have an egg around. If you have really regular periods, you keep track of them carefully and you can estimate the time when you ovulate by changes in the thickness of cervical mucus or body temperature, you might have a slightly better chance to avoid pregnancy for a while. But even when you are really careful about only having sex at 'safe' times, it is not a very effective method for preventing pregnancy. Timing your sex leads to pregnancies more often than using birth-control pills, condoms or injectable hormones. If you have sex without any of these other kinds of birth control, there is really no safe time to have sex. If you don't want to get pregnant, use real birth control. And if you don't want to get a sexually transmitted disease, you also need to use a condom. (See, we really do love condoms.)

Myth: You can't get pregnant if you have sex in water

You can get pregnant if you have unprotected sex, no matter where you do it.

If a penis is in a vagina, then it really doesn't matter what's going on anywhere else. Warm water, cold water, chemically treated water – none of these will make a difference. You are just as likely to get pregnant during sex in water as during sex out of water. Granted, if a man ejaculates into the water, it is pretty unlikely that the sperm will find their way into the vagina and up to the egg. Moreover, if the water is hot enough or treated with enough chemicals, the chances are much better that the sperm will die off long before they even have a chance to get close to the vagina.

But never say never. If there are living sperm and living eggs around and some way for them to get together (no matter how unlikely), you can get pregnant.

Myth: Birth-control pills don't work as well if you're on antibiotics

If you are using birth-control pills, you've probably been advised that they will be less effective if you are simultaneously taking an antibiotic. Some people suggest also using condoms to protect against pregnancy if you need to take antibiotics.

Despite widespread fears about this (and the warning labels you may see on some prescriptions), there is no good science to suggest that birth-control pills don't work as well while taking antibiotics. One review by the American Academy of Family Physicians concluded that the scientific literature does not suggest that common antibiotics reduce how well birth-control pills work. Some birth-control pills have a low hormone dose to prevent pregnancy, and some of these might work less well when combined with antibiotics; but again, the difference in efficacy is minimal. Another study looked at 356 patients in three dermatology practices with a history of long-term use of antibiotics and birth-control pills together. There was no statistically significant difference between how many women got pregnant in the group on both antibiotics and birth-control pills and the control groups where women were taking just the birth-control pill. Remember, birth-control pills fail at least 1 per cent of the time even in ideal conditions. And in studies that look at what happens in real life when women take an antibiotic with their birth-control pill, the rate of getting pregnant doesn't seem to change.

Some science does suggest that there is a possibility of one antibiotic, rifampin (a drug usually used to treat tuberculosis), making birth-control pills less effective. In a study of thirty women, the level of the hormone in the birth-control pill that prevents you from ovulating was lower in the women's blood when rifampin was also being taken. However, none of the women in the study became pregnant as a result of taking rifampin and birth-control pills at the same time.

Future research of new drugs and rare antibiotics may discover that you should not take antibiotics and birth control at the same time, but current science suggests that this is rarely a concern. It is much more important to take your birth-control pill every day at the same time than to spend time worrying about your antibiotics.

Myth: You can't get pregnant when you're on the pill

Birth-control pills work a lot better to prevent pregnancy than any of the other methods we have talked about. Much, much better than standing up. Much better than trying to pull out. Much better than restricting your activities to the hot tub. But the pill is not perfect either. Again, if there is a sperm and an egg involved, pregnancy is always a possibility.

In the course of a year, between 5 and 8 out of 100 women using the pill will have an accidental pregnancy. Those are better odds than for other methods (remember, it was 27 out of 100 women getting pregnant using the withdrawal method), but it's not down to zero. Birth-control pills work best when a woman takes them every single day at the exact same time of day (especially if you are on an oestrogen-only pill). If you take the pills absolutely perfectly, you are even less likely to get pregnant. But if you are not good at taking a pill every day and at the same time every day, your chance of getting pregnant while using the pill is higher. Missing even one pill can significantly increase your chance of getting pregnant.

Myth: You're more likely to have a baby boy if you conceive in the middle of your cycle

If you have a brood of girls and you're trying one last time to have a boy, you've probably heard a lot of urban myths about how to increase your chances of having a boy. Before we even look at the studies, just remember that the sex of your baby is determined solely by the sperm. If the sperm has a Y chromosome, then the baby will be a boy. If the sperm has an X chromosome, then the baby will be a girl. What it comes down to is how (and if) you can help the sperm with the Y chromosome win the race to fertilize the egg.

While your grandmother or that nice lady down the street might have some ideas about sure-fire ways to have a boy, scientists have really only debated about one method of sex selection in particular. Several studies have looked at whether there is a connection between the sex of the baby and the time of the month when the egg is fertilized. In particular, scientists have investigated whether the sex of the baby can be determined based on the time between when you have sex and when you ovulate. Researchers suggested that you could increase your chance of having a boy if you had sex very close to the time when the woman ovulated. The mucus from a woman's cervix changes during a woman's cycle, with the mucus getting thicker during the middle of the cycle, around the time of ovulation. They theorized that you might be more likely to get boys when the mucus is thicker (close to the time of ovulation) because sperm with a Y chromosome are more mobile and thus more likely to penetrate that mucus than the sperm with an X chromosome. And some of the studies testing out this theory reported slightly more boys conceived during sex close to ovulation, enough to show a small but statistically significant difference in whether you get girls or boys from sex at different times of the month.

Unfortunately, the more scientists study this, the less consistent the results. In a meta-analysis that combined all of the data from six studies that looked at the timing of intercourse and whether you ended up with a boy or a girl, researchers actually found fewer boys among those conceived during the most fertile time of the cycle (right around ovulation). The differences were still small, but completely the opposite of the aforementioned findings. In another study of the sex-selection technique involving seventy-three women over four years, there was no difference in the number of boys or girls among those who tried to select the sex of the baby by timing their intercourse versus those who did not.

Myth: You can predict the sex of your baby without a doctor

Pregnant bellies inevitably invite commentary by family, friends and complete strangers. If you have not found out the sex of your baby by ultrasound or by laboratory tests, people are particularly quick to offer their predictions. The methods of predicting the sex of the baby are endless. Are you carrying high or carrying low? Does your abdomen look like a basketball? How bad is your morning sickness? How does this pregnancy compare to previous pregnancies? What does your intuition say? How much is the baby moving around? Other people recommend particular tests to predict the baby's sex: Is the baby's heart rate low or high?

The truth is, your chance of randomly guessing the sex of your child is about fifty–fifty. And, as it turns out, it is pretty difficult for any bogus method to beat those odds. Doctors and scientists alike have put this folklore to the test, and it turns out that these methods are pretty much bunk:

Weight gain and shape of the belly: In a study of 104 pregnant women who did not know the sex of their baby before giving birth, the shape of their bellies during pregnancy was not an accurate predictor of the baby's sex. In another study of 500 births, neither the mother's weight nor her weight gain during pregnancy helped determine whether the baby was a boy or a girl.

Heartburn: In a study that measured how much heartburn sixty-four pregnant women experienced during pregnancy, the severity of the heartburn symptoms had absolutely nothing to do with the sex of the baby. This study did have one interesting finding; the more severe a woman's heartburn, the fuller the head of hair on the newborn.

Mother's intuition: In a study of 212 pregnant women, 110 patients had a strong feeling whether the baby was a girl or a boy, but maternal intuition was no better than random guessing for predicting the baby's sex. In another study of 104 women, 55 per cent of the women correctly guessed the sex of the baby, but statistical analysis shows this again to be no better than random guessing. Furthermore, mothers did not do any better whether they made the prediction early, late or in the middle of the pregnancy.

Baby's heart rate: Some people claim that if the foetal heart rate is 140 beats per minute or faster the baby is a girl. If the foetal heart rate is 139 beats per minute or lower, then it's a boy. However, scientific data suggests that there is no significant difference in the baseline foetal heart rates of a male and female foetus at any recorded gestational age.

Morning sickness: In cases of hyperemesis gravidarum – the very worst kind of morning sickness, which is defined as excessive, unrelenting nausea and/or vomiting that prevents a pregnant woman from taking in enough food or fluids – you may have a better chance of having a girl than a boy. Several studies support this finding. However, the difference isn't huge, so even if you are puking your guts out all the time, the little one in your belly may very well be a boy.

We recommend another test for you that works about as well as all of these – try flipping a coin!

Myth: Twins skip a generation

At best, the idea of twins skipping generations is a half-truth. Twins can run in families. Scientists have found a gene that predisposes some women to release more than one egg in a single cycle of ovulation. If both of those eggs are fertilized, then this 'hyperovulation' can lead to fraternal twins. Genetically, fraternal twins have just as much in common as any two siblings, but they happen to share a uterus and to be the same age.

Parents pass genes on to their children, so if they pass on a gene that leads to hyperovulation, the girl who got that gene would be more likely to have twins. However, men obviously are not affected by the hyperovulation gene, so if the gene is passed to a male child, it's useless in terms of his own child-bearing. If he has a daughter, though, and passes the hyperovulation gene to her, then she may be more likely to have twins. In this very contrived scenario, twins would skip a generation. However, since all of this relies on the gene eventually getting passed down to a female, there is no guarantee that twins always skip a generation.

Furthermore, not all types of twins run in families. Identical twins are the result of one fertilized egg that then splits into two, which creates two babies who share the exact same DNA. Scientists do not know of any genes that influence whether a fertilized egg splits into two – it seems to be a random occurrence. If your extended family has several sets of identical twins, this is probably just a coincidence.

Myth: Flying on a plane is dangerous for your unborn baby

Unless a pregnant woman has a specific medical problem or problems with her pregnancy, the American College of Obstetricians and Gynecologists, the biggest group of OB/GYN doctors in the US, states that pregnant women can fly safely up to thirty-six weeks of gestation. The climate of the aeroplane, including elements like the low humidity in the cabin and changes in the pressure in the cabin, do temporarily change the mother's heart rate, blood pressure and breathing, but they have not been proven to have any detrimental effects on the baby. There are some unconfirmed reports that flight attendants have an increased chance of miscarrying or having a spontaneous abortion compared to non-employed women, but not compared to other employed women. However, there is no scientific evidence of an increased risk of miscarriage among other pregnant women who fly.

One study followed 222 women, of whom 118 travelled by air at least once during their pregnancy. When the doctors compared the two groups, there were no differences in length of pregnancy, the risk of having a premature baby, the babies' birth weights, the risk of vaginal bleeding, how often the babies were admitted to the neonatal intensive care unit, or in any combination of all the possible things that could go wrong during pregnancy.

A major concern with lengthy air travel is deep-vein thrombosis (DVTs) or clots in the veins of the legs that can then travel to the lungs. Experts suggest that pregnant women may be at increased risk for having DVTs, but there are no published reports of an actual increase in DVTs in pregnant women because of air travel. Nonetheless, all air travellers should take precautions to avoid DVTs by frequently flexing the ankles and feet and taking regular walks in the aircraft.

Some pregnant women also fear exposure to noise vibration or to cosmic radiation in the atmosphere while travelling by airplane. There is not a lot of scientific evidence that has tested whether these are problems for an unborn baby. The existing studies of the health effects of aircraft noise and of galactic cosmic radiation exposure for anyone during air travel indicate that any potential risk to a pregnant woman is so small that it should not be cause for alarm.

Myth: Bed rest prevents pre-term labour

The widespread practice of bed rest is based on the common-sense theory that strenuous work or play may make a woman have contractions and go into labour prematurely. However, the best (and only) answer to the question of whether bed rest actually works for pre-term labour comes from four researchers who conducted a systematic review of the literature, looking for studies investigating what happened to women at high risk of giving birth prematurely and to their babies after the pregnant women were put on bed rest. The review authors could only find one study that really investigated this question. It was a large study of 1,266 women, and it showed that bed rest did not prevent pre-term births.

We recognize that this assertion is going to be met with a lot of angry responses. However, the authors of the systematic review concluded, as do we, that there is *no evidence* to support bed rest to prevent pre-term birth. We have *no evidence* that it works, and we have one pretty large study in which it didn't seem to work. While we can't prove that bed rest does not help certain women, we are far from being able to prove that it does help.

Doctors and pregnant women afraid of pre-term labour also need to remember that bed rest is not necessarily completely harmless; it can cause problems like deconditioning of the muscles. In addition, unnecessarily preventing mothers from working can create significant problems for the family's finances, and even for society as a whole.

PART IV

'He won't get into Oxbridge without Baby Einstein®'

Myths About Babies and Children

Myth: Baby Einstein® makes your baby smarter

While wandering down toyshop aisles full of educational toys, CDs and DVDs that promise to maximize a baby's brain development, many parents feel guilty if they do not buy products that are specially designed to make their babies smarter. Parents spend billions of pounds every year on Baby Einstein®, Brainy Baby® and other products like these. The creators of these products claim that they are 'specifically designed to promote discovery and to inspire new ways for parents and babies to interact'. The references to 'Einstein' and 'the Mozart effect' lead parents to believe that these products really will give their child a better shot at becoming a prodigy.

Research suggests that popular baby DVDs do not, in fact, create geniuses. In fact, they may even slow a baby's development. A research team from the University of Washington surveyed 1,008 parents of children aged between two and twenty-four months, asking about the TV programmes and DVDs their children were watching. The parents also completed an evaluation of their child's language development. In infants aged between eight and sixteen months, each hour per day of viewing baby DVDs or videos was associated with a significant drop in their language development skills. The babies watching these videos scored lower in assessment of their language skills than babies who weren't exposed to the DVDs. For toddlers, the effects of the baby DVDs and videos weren't quite as worrisome. While watching these DVDs was still not associated with any improved language development, neither was it associated with any drop in their language development.

They may not be turning into baby Einsteins, but what about

when they get older? Looking at older kids, one study suggests that watching television before the age of three is associated with decreased mental development at ages six and seven, as evidenced by lower scores on tests of reading recognition, reading comprehension and remembering numbers. In a compilation of twelve studies that examined the effect of television content on children aged three and older, educational television programmes were found to be successful in broadening children's knowledge. In the short term, educational programmes might change children's attitudes about particular ideas. For instance, in one study, showing children television segments that featured non-white children playing made the children more likely to select a photograph of a non-white child when they were asked to select photographs of children they would like to play with. However, in the studies of television-watching in this review, there was no convincing evidence that TV affects children's prosocial or aggressive behaviour.

Myth: Adding cereal to your baby's diet helps him sleep longer

There is a good reason why the proverb 'never wake a sleeping baby' has remained in use for so many years. It's because babies who don't sleep turn their parents into exhausted, delusional, raving lunatics who bear only a slight resemblance to their pre-parental selves. When your baby is still demanding that 2 a.m. feed, and someone suggests adding a bit of cereal to his diet, either on a spoon or straight into the bottle, the chances are that you will be more than happy to try it.

Both as paediatricians and as parents, we have heard from hundreds of people that cereal helps babies sleep. Paediatricians usually recommend starting babies on cereal at no earlier than four months old. The belief in the sleep-inducing powers of cereal may cause parents to disregard this recommendation and start solids earlier than recommended. In a survey of 102 four-month-old children, 44 per cent of mothers introduced solids before four months of age. 80 per cent of these mothers said that their child was not satisfied with formula or breast milk alone, and 53 per cent said that the cereal helped their child to sleep better at night.

Is there any proof that adding cereal to babies' milk does actually help them sleep? Over the last thirty years, scientific studies have shown that cereal does not help babies sleep longer. In 1974, the *British Medical Journal* published a study of 100 mothers and babies that showed no link between the introduction of cereal and whether the babies slept through the night. A study from 1989 randomized 106 infants to have cereal in their bedtime bottle at five weeks or at four months of age. The parents recorded their baby's sleep patterns every week from four to twenty-one weeks of age. Babies who got cereal in their bottles were no more likely to sleep through the night (whether defined as six hours or eight hours in a row), nor were they more likely to sleep longer. The cereal just doesn't work.

If cereal does not work, is there any change in diet or routine that parents can implement to help their babies sleep through the night? One study showed that babies who had more than eleven feeds per day at one week old were at a higher risk for sleep problems when they were twelve weeks old. This same study showed that behavioural programmes to help parents modify their baby's eating and sleeping patterns helped the babies sleep through the night at a younger age. The structured behavioural programme that was studied included setting distinctions between day- and night-time environments, and delaying a feed when the baby woke at night (e.g. by changing the baby's nappy first). This strategy is further supported in other studies, where babies were found to sleep through the night earlier when their caregivers adopted a system of feeding the baby between 10 p.m. and midnight every night, and gradually lengthening the time between middle-of-the-night feeds by doing other things to care for the baby (experts recommend re-swaddling, changing the nappy or walking the baby, for instance). In addition, parents are encouraged to emphasize the differences in a baby's daytime and evening environments. Babies whose parents tried these changes in diet and routine slept through the night earlier and fed less frequently at night – without the help of cereal.

Myth: Propping a baby upright helps with reflux

Some babies are big pukers. They routinely cover their parents' shirts, trousers, necks, faces and any other available surface with the uniquely unpleasant curdled and sour liquid that is baby sick. Some babies are such prolific pukers that their parents and grandparents wonder how they can possibly gain weight since they are not keeping down any breast milk or formula at all. When the parents of these babies drag themselves into the paediatrician's office, baby and parents alike in stained and discoloured clothing, everyone involved wants to do something to fix the problem.

One of the solutions usually suggested is to keep the baby upright after he or she eats. Positioning the baby upright is thought to aid gravity to keep that milk in the baby's stomach, easing the pressure or preventing it from squirting back out. Some parents follow a baby version of the adult recommendation to prop up the head of the bed, and they tilt the cot or even buy a special wedge to help keep the baby in a more upright position while sleeping. Much effort goes into finding ways to keep the baby tilted upright (in addition to all the clothes-washing and carpet-cleaning).

Unfortunately for the parents of pukers, and for anybody else unfortunate enough to hold one of these little time bombs, the evidence tells us that positioning a baby upright does not ease their reflux. Two systematic reviews (one of them done by Aaron himself!), studies that look for and compile every single study related to a particular question, both conclude that positioning does not help reflux in children under two. Five studies examined whether upright positioning decreases reflux in babies, using a very sensitive probe that is actually placed in the baby's oesophagus to measure changes in acidity and provide a very precise measure of whether or not any reflux is taking place, even if the baby does not vomit. Elevating the head of the bed or positioning the baby upright after meals did not have any significant effect on the reflux of the babies studied. Positioning the baby at 60° elevation in an infant seat was actually found to increase reflux compared to positioning the baby in a prone position on his or her back.

The best solution of all for reflux is the passage of time, as most babies grow out of their reflux as they get older and the sphincter around their oesophagus gets tighter. Until then, you may want to think about a laundry service.

Myth: Teething can cause a fever

Teething can cause many symptoms, but fever is the one we care most about. Whether teething causes a fever can have real implications for a child's health and their healthcare. Doctors take fever in infants very seriously. If a baby has a fever and it is not clear why, we may have to test their blood or urine or maybe even put a needle in their back (otherwise known as a lumbar puncture or spinal tap) to find the source of the infection. If the fever can be chalked up to teething, then doctors might not need to be so worried about looking for other causes of the fever. In a survey of 462 parents, nurses and paediatricians, almost all of the parents surveyed reported that they thought teething caused a fever. In a similar survey of paediatricians, paediatric dentists and parents, the majority of each of these groups believed the same thing.

Because plenty of people obviously share the same concerns, we wanted to find out the truth – which we uncovered in two separate studies. The first study followed 123 teething children aged between four months and one year. In this study, parents took their child's temperature twice a day, and recorded whether they had any of the following symptoms: increased biting, drooling, gum-rubbing, sucking, irritability, wakefulness, ear-rubbing, facial rash, decreased appetite for solid foods and mild temperature elevation. In addition, parents kept track of every tooth eruption. The parents recorded more symptoms in the four days before a tooth emerged, the day of the emergence and the three days after the tooth popped through – in total, an eight-day teething period. Despite experiencing any combination of the aforementioned symptoms, a fever above 102° Fahrenheit was not found to be commonly associated with teething.

In fact, more than 35 per cent of the teething infants had no symptoms at all – and no single child in the study experienced a fever of 104° Fahrenheit or a life-threatening illness of any kind. A second study, using similar daily temperature and symptom records, found teething was not associated with either a 'low fever' (a temperature below 102° but above 98°) or a 'high fever' (102° and over).

Although teething can certainly make a child uncomfortable and may cause some changes in a baby's behaviour, fevers should not be explained away by teething. There is no evidence to support a clear connection between fever and teething, and parents and doctors alike should look seriously for other causes of fever in a teething baby.

Myth: It's possible to spoil a baby

The fear of spoiling the baby leaves many new parents questioning their instincts and wondering whether they should really hurry over to pick up their screaming infant. In a survey of 531 parents, the majority of mothers and fathers believed that a baby could be spoiled, even though people had different ideas of just what this spoiling would look like and what the future effects might be. In that study, younger or less-well-educated parents were more likely to have rigid ideas about just what will spoil a baby and about the negative effects that might result. In another study, 129 women were questioned about their social support, their perception of spoiling babies and whether they had any symptoms of depression. Fifty-eight per cent of the mothers believed that infants less than five months old could be spoiled. Those who believed that a baby could be spoiled were more likely to be first-time mothers and more likely to have been depressed during their pregnancy.

Do these parents have it right? There are certainly plenty of demanding kids with poor manners, not to mention self-important adults who seem to think the world revolves around them. Is this because they had someone rushing to meet their every need from the moment they were born? Not necessarily. The expert opinion is that you cannot spoil a young baby. New babies need as much love and care and attention as you can give them, and they are too young to start being independent. In other words, they need a lot of 'spoiling'. Whereas 'spoiled children' might use crying to manipulate you into giving them whatever they want, babies are too young to deliberately manipulate you. Young babies cry to let you know that they need something. It's the only way they can communicate. And a quick response actually builds their trust. Paying attention to the baby right away can actually help them to feel more secure, less anxious and less likely to cry for no reason at all. Thus, in the long run, responding to the baby right away should help them to be less demanding and clingy.

The problem with believing that you might spoil your baby is that it might lead you to misunderstand your baby's basic need to be nurtured. When this happens, when there is a mismatch between what the baby needs and what the mother provides, a baby might become less secure and trusting. Furthermore, one of the keys to healthy attachment between a baby and the mother seems to be how quickly the mother responds to the baby, more significant than the time spent comforting the baby.

Once a baby is six to eight months old, however, he or she begins to explore cause and effect and to make connections between what they do and how you respond. They might want to drop a toy over and over again, for example, to see it fall and then to get you to pick it up. This is the time when it is more appropriate to set limits on what you do or don't do for your baby. While a child should get plenty of hugs and praise for good behaviour (and at other times in between), you might try to hold out a bit when a child is crying for something that is not really needed. Then you can rush back in with the hugs as soon as your child is calming down. In the second half of a child's first year, children reach a time in their development when they still need to trust their parents, but they also need to learn to trust themselves and to be a little more independent when it comes to things that they want (but don't need). This is when you might change the limits you set as a parent, helping your child one step further on the road to independence.

Myth: Women who are breastfeeding can drink alcohol

Some women think that it is okay to drink alcohol as long as they drink during their breastfeeding, so that the alcohol does not have a chance to get into the milk. Others think that they can have a drink and then pump and chuck the milk that follows. Others have heard that if they have a drink and then wait an hour before feeding, everything will be fine. Still others, especially in cultures where alcohol is thought to increase milk production, believe that alcohol is not a problem for a breastfeeding baby or that it does not even get into the breast milk.

Drinking alcohol is relatively common among breastfeeding women; in a survey of 772 breastfeeding women and 776 non-breastfeeding women, 36 per cent of the breastfeeding women consumed alcohol as early as three months after giving birth. Although they were less likely than non-breastfeeding women to have two or more drinks per week, they were just as likely as those women not breastfeeding to have one drink per week.

Let's set one thing straight – alcohol does enter breast milk. Alcohol actually concentrates in breast milk and can pass directly to the baby, and this has been shown by numerous studies testing breast milk for alcohol content. The amount of alcohol that reaches the baby is small, but it is there nonetheless – and in larger amounts (more than the equivalent of a small glass of wine), alcohol has been shown to inhibit a mother's milk production.

The timing of drinking and breastfeeding is a little more tricky than you might think. When the alcohol is in your bloodstream, it is also in the milk that you produce. Drinking just before you nurse, or during nursing, may not help matters at all. In order for the alcohol to completely leave your bloodstream, you should probably wait to breastfeed for at least an hour after your last drink.

In the short term, alcohol in the breast milk changes the behaviour of the baby as he or she feeds, and babies actually take in significantly less milk if it contains alcohol. This has been shown in a small study in which babies were videotaped as they nursed, both after their mother drank plain orange juice and after she drank orange juice with a small amount of alcohol in it. A panel of adults judged the smell of the breast milk both with and without alcohol (they thought the alcohol-laden milk smelled different), and the samples were analysed to see how much alcohol they contained. And while you might think that having an alcoholic drink before you nurse will help the baby to sleep better (perhaps because a glass of wine makes you more sleepy), the reverse is actually true. In a different study, babies slept significantly less and for shorter periods of time after their mothers drank alcohol.

Aside from taking in less milk and becoming a bit restless, you may wonder if there are any more serious side effects for a baby from having milk that contains a little bit of alcohol. The effects of alcohol from breast milk on a child's development are not completely clear. Some studies have shown that alcohol in breast milk is tied to decreased motor development in babies. In a study of 400 infants, those who ingested breast milk that contained alcohol had lower scores on a test of motor development (measuring things like crawling and walking) when they were one year old, but they did not score significantly lower on tests of mental development. Thirty of the 400 infants were considered to have had a high infant alcohol score. In a subsequent study of 915 eighteen-month-old toddlers, researchers did not find a link between exposure to alcohol and delayed motor development. However, they cautioned that we don't know whether there are any effects later in life.

Myth: Over-the-counter cold and cough medicines are safe for babies and toddlers

Coughing, sneezing, runny noses, whining, crying. When kids have colds, it is no fun for anyone. When your baby is up all night fussing, crying, coughing and stuffed-up, you would probably do anything to make them feel better. You've probably seen the countless cough and cold remedies in the shops, complete with smiling babies on the box.

Because young children have on average ten to twelve colds or viruses a year, it can seem as though they are ill all the time. As paediatricians, we really wish that we could recommend something to help parents and children feel better. Unfortunately, over-the-counter cough and cold medicines are not the answer. First of all, they don't work. Since 1985, there have been six randomized, placebo-controlled studies of the use of cough and cold medicines for children under twelve years old. All six showed that there was no difference between taking a cough or cold medicine or taking a placebo or fake medicine. Expert panels have reviewed these studies and have all agreed that there is no evidence to suggest these medicines work for kids. The American College of Chest Physicians stated that the scientific literature for over-the-counter cough medicines did not support that cough medicines worked. A 1997 policy statement from the American Academy of Pediatrics stated that support for the use of cough medicines in children had not been established. Neither the medical experts, nor the scientists, nor the chest doctors, nor the paediatricians think that these medicines work.

Even worse than not working, over-the-counter cough and cold medicines have bad side effects and can even kill children. In 2004 and 2005, 1,519 children in the United States under the age of two years had to go to A&E because of a bad reaction to or an overdose of cough and cold medications. The bad reactions included serious problems like abnormal heart rhythms, loss of consciousness and brain damage. In 2005, three infants under the age of six months died because of cough and cold medicines (this was verified as the cause of death by the medical examiners or coroners in each of these instances). A US Food and Drug Administration review found 123 deaths that could be tied to cough and cold medicines used in children under the age of six years over the past thirty years.

The bottom line is this: if something doesn't work well *and* it has the potential to put your child in hospital or kill them, we don't think it's a good idea.

After many years of doctors and scientists stating that these cough and cold medicines don't work in children, the FDA is now publicly issuing stricter statements against them, causing many to be pulled from the shelves. Following the recommendation of the FDA, the UK Medicine and Healthcare Products Regulatory Agency (MHRA) recommended that over-the-counter cough medicines were not suitable for children under two years of age. We hope that these statements and recalls help more parents recognize the danger of these medicines.

Myth: Walkers help your baby learn to walk earlier

Parents seem to love walkers because they imagine that they might give their child a headstart in learning to walk, plus they keep the baby entertained. However, contrary to parental expectations, baby walkers may actually delay a child's ability to walk by herself.

Studies have shown that infants who used baby walkers did not learn to walk until two to four weeks later than infants who did not use walkers. Additionally, two comprehensive reviews, compiling the results from nineteen studies of baby walkers, found that using baby walkers delayed the motor development of infants. Two studies found delays in walking of between eleven and twenty-six days when babies had been using walkers. In a particularly concerning study of 109 infants, those who used baby walkers sat, crawled and walked later than those with no walker experience, and they also scored lower on tests of mental and motor development. Two other randomized, controlled trials did not show a delay in when babies learned to walk if they had been using walkers, but they showed no advances either. The take-home message: walkers do not help babies walk earlier, and may even cause them to walk later.

Why do walkers possibly cause these delays, when babies really do look like they're learning to walk when they use them? First of all, walkers usually block babies' ability to see their legs moving – which is crucial in their ability to learn to walk. Second, babies walk differently in baby walkers than they do when they walk on their own. In a walker, they use the muscles at the back of their legs and often walk on tiptoes, causing their muscles to develop and tighten improperly. To crawl or walk properly, babies need to use the muscles at the front *and* the back of their legs. Babies who learn to walk in walkers need to relearn how to work these muscles before they can walk on their own. Finally, walkers provide balance for a baby, which means they are not prepared for the challenges of balancing themselves when they walk unaided.

Still, some parents just like to use walkers to keep their baby entertained. However, these parents should be careful – baby walkers can easily fall down the stairs, into other dangerous places like fireplaces or pools, and even roll down a driveway into the road. Baby walkers commonly cause head injuries, broken arms and legs and facial injuries. If you have a baby walker in your house, your child is nine times more likely to end up in the A&E department with an injury. In surveys, between 12 and 50 per cent of parents whose babies use walkers report that their child has had some sort of walker-related injury. Some estimates say that more than 20,000 infants are injured in the United States every year because of baby walkers.

Since walkers don't work particularly well, and they have the potential to harm your child, we advise you to stick the walker in the attic and forget about it.

Myth: The iron in baby formula causes constipation

When a baby is straining and turning red and struggling to produce poos like hard rabbit droppings, solving the constipation problem becomes a high priority. Well-meaning people might suggest that the iron in the formula milk may be to blame, and parents sometimes think that switching to a low-iron formula will help. No one wants pellet-like poos, but is iron-fortified formula really the culprit?

No.

Researchers in Britain conducted a randomized, controlled trial of 493 infants who were given either cow's milk, iron-fortified formula or low-iron formula. Over a period of nine months, they carefully followed any problems the infants had with infections, diarrhoea or constipation. Infants who drank the iron-fortified formula did not have any more constipation (or any more diarrhoea or other infections) than the babies who drank low-iron formula or cow's milk. Moreover, in a study of 300 children, half of whom were given iron supplements and half of whom were given a placebo or fake medicine, the infants taking the placebos had more constipation than the babies taking the iron. There is just no evidence that iron causes constipation in children.

In case you are now worried that iron might instead cause diarrhoea, a different randomized, controlled trial of 1,655 infants found that iron-fortified formula did not lead to any more diarrhoea than using low-iron formula.

Formula with iron also provides important benefits to babies. Not having enough iron can lead to anaemia (red-blood-cell deficiency). Infants who receive formula fortified with iron have significantly less anaemia than infants who are given low-iron formula or plain cow's milk. Severe anaemia can even lead to developmental delays or heart failure.

The evidence is fairly conclusive that iron-fortified formula does not give babies constipation, and that without enough iron in their diets, babies face significant health risks. That should be reason enough to make sure that you give your baby formula with full iron until they are one year old. If your baby keeps pooing pellets, talk to your doctor about other possible causes of bowel distress instead of switching the formula.

Myth: Babies need water when it is hot outside

Just as many adults believe that they should drink eight glasses of water a day, many believe that babies need water too. Especially when it is hot outside, parents and those who advise parents often give babies water in addition to their formula or breast milk. In a survey of 642 mothers of eight- to nine-week-old infants, 61 per cent said their child drank some quantity of water and in amounts much higher than the intake recommended for infants younger than one year by the US Environmental Protection Agency.

While everyone needs to be hydrated, babies usually do not need any 'extra' water. Babies get the water that they need from their breast milk or formula. There are no studies to support the need to give infants any additional water. In contrast, the American Academy of Pediatrics specifically states that 'until your baby starts eating solid foods, he will get all the water he needs from breast milk or formula'.

While there are no studies to support the need to give water to babies, there are studies that report some very real dangers of giving babies water. If infants get too much water, they are at risk of having too low a concentration of the normal electrolytes in their blood. A baby's developing kidneys may not be able to get rid of the water quickly enough, leading to a build-up of water in the body which dilutes the sodium balance of the blood. This condition is called hyponatremia or water intoxication, and can result in seizures, coma and even death. Cases of hyponatremic seizures have been reported after parents fed babies commercial bottled water that is explicitly marketed for use in infants (sometimes called 'nursery water'). Babies are even susceptible to hyponatremia if their parents do not mix their formula properly and add too much water to it. If your baby is throwing up and having diarrhoea, don't just give them water.

Children who are ill may need extra fluids, but water is usually not the best choice. Oral rehydration solutions are better, and this is a good reason to consult your doctor if your baby or young child has a stomach bug. If you sense that your baby is thirstier on hot days, a little extra breast milk or formula (instead of water) will better help them to get the fluid balance they need without risking water intoxication.

These potential problems with too much water for babies are also why paediatricians are not necessarily big fans of early 'swimming lessons' for babies. 'Water babies' who swallow too much water during their 'swimming lessons' have developed hyponatremia and seizures.

Myth: Sugar makes kids hyperactive

Mike Myers once did a *Saturday Night Live* sketch in which his character, Simon, a boy with ADHD, ripped a climbing frame from the ground and dragged it across a field after eating some chocolate. As funny as that scene was, the truth is that sugar doesn't really make kids hyper.

There have been at least twelve trials of various diets investigating different levels of sugar in what children eat. None of these studies detected any differences in behaviour between the children who had eaten sugar and those who hadn't. These studies included sugar from sweets, chocolate and natural sources. Even in studies that included only children who were considered 'sensitive' to sugar, the children did not behave differently with a diet full of sugar or one that was sugar-free.

This myth, perhaps more than any other, is met with disbelief when we discuss it, especially with parents. Scientists have even studied how parents react to the sugar myth. In one well-thought-out study, children were divided into two groups. One group was given drinks that their parents were told were sugar-free. The other group was given drinks that the parents were told were full of sugar. Then all the parents were told to grade their children's behaviour. Not surprisingly, the parents of children who thought their children had drunk a ton of sugar rated their children as significantly more hyperactive. The twist to this study was that all the children were given the same drink, sweetened with sugar-free aspartame, so the differences in behaviour were all in the parents' minds.

Even when science shows time and again that this is not the case, we continue to persist in believing that sugar causes our kids to be hyperactive. It's just not true. This does not mean that there are not very good reasons why your kid should not ingest large quantities of sugar. As almost any parent knows, sugar has been linked to tooth decay and increased weight gain – just don't blame it for your child's bad behaviour.

Myth: Eat your spinach to grow strong like Popeye

Kids dread the green, leafy spinach on their dinner plates, pushing it around in circles in the hope that no one will notice that they are not eating it. Nonetheless, parents hold firm in their conviction that spinach is great for their kids, that it will help them grow strong (like Popeye, they say persuasively). Celebrity chefs have even come up with ways to sneak spinach into other foods that children enjoy, like brownies or muffins. After all, spinach is such a great source of iron that it must be worth getting kids to eat it any way you can.

We (well, really just Rachel) love spinach in many forms – in salads, cooked, in quiches – but once again, we need to bring a bit of science to the claims about the strength-giving powers of spinach. (Feel free to ban your children from reading this section of the book. Understanding the truth about spinach ends up disappointing fans of the sailor man.)

Spinach is routinely recommended as a good source of iron. Spinach does contain non-haem iron, the type of iron found in vegetables. For a vegetable, spinach is a relatively good source of iron. There are 1.9 milligrams of iron in a 60-gram serving of boiled spinach, and slightly more if you eat it raw. Most green vegetables have less than 1 milligram of iron in a serving. In comparison with foods that come from animals, though, spinach contains a relatively small amount of iron. If spinach does not really have all that much iron, how did it get this great reputation?

The buzz about spinach may have started back in 1870 when Dr E. von Wolf published a paper reporting that spinach had a remarkably high iron content. Unfortunately, Dr von Wolf did not have a good copy editor, and he ended up with a decimal point in the wrong place in his numbers. So he accidentally reported an iron-content figure for spinach that was ten times too high. It was probably because of these mistaken findings that spinach became Popeye's miracle food in the 1930s. As described by Dr T. J. Hamblin in the *British Medical Journal* in 1981, some German chemists in 1937 (perhaps wanting to prove that Popeye had the wrong idea) decided to investigate the wonder vegetable and realized that Dr von Wolf's figures were wrong.

One might think that it is good enough that spinach has more iron than most vegetables. Surely spinach will still help you to build up your red blood cells and muscles. Unfortunately, the body has a harder time using non-haem iron, the type of iron found in spinach. It is generally absorbed pretty slowly. There are chemicals, like Vitamin C, which can help the body absorb this type of iron better. But there are also chemicals which make it even more difficult for the body to absorb this type of iron. And ironically, spinach actually contains one of these chemicals. Oxalate, a chemical that binds with iron and prevents it from being absorbed well, is found in spinach.

Another claim for spinach as a wonder vegetable is its high calcium content. Unfortunately, the oxalate in spinach causes problems for absorbing calcium too. Oxalate also binds to calcium and makes it harder for your body to absorb it. For example, your body can absorb 50 per cent of the calcium in broccoli, but it can only absorb about 5 per cent of the calcium in spinach.

In the end, you can force your kids to eat all the spinach that you want, but it is unlikely to have dramatic effects on their health and strength. Spinach is a great vegetable; it just doesn't deserve any extra credit for providing lots of iron or calcium.

PART V

'Don't swallow your gum!'

Myths About What We Eat and Drink

Myth: Chewing gum stays in your stomach for seven years

Don't swallow your gum! You must remember being warned as a child not to swallow that piece of chewing gum. Even though swallowing it seemed so much easier than spitting it out and finding a proper place to throw it away, you always knew that gum might stay in your stomach for seven years. That's enough to make anyone pause before they let that piece of gum slide down their throats.

Almost every honest gum-chewer would admit to swallowing a piece of gum every now and then. What happens to all of that gum? Is it really sitting around in your stomach for seven years?

Chewing gum is made out of gum base, sweeteners, colouring and flavouring. The gum base is pretty indigestible stuff; it is a mixture of elastomers, resins, fats, emulsifiers and waxes (sounds delicious, right?). Most of the time, your stomach really can't break down the gum the way it would break down other foods. Sugar and other sweeteners in gum might get absorbed, but those waxes and resins and elastomers resist the powers of your stomach juices to break them up. However, your digestive system has another way to deal with things you swallow. After all, we eat lots of things that we can't fully digest. The gut just keeps moving them along until they make it all the way through the intestines and come out the other end. So gum usually ends up in your toilet one to two days later, pooed out by the power of peristalsis, which propels material through your bowels. Even though gum is sticky, it is usually no match for the power of your gut.

The system can get gummed up though (ha ha). Swallowing a huge wad of gum or swallowing many small pieces of gum in a short time can cause a blockage within the digestive system (most often in children, who have a smaller-diameter digestive tract than adults do), but this is extremely rare. Kids are more likely to get blocked up by gum if they swallow the gum along with other things that can get stuck with it (like coins or sunflower seeds).

While we don't recommend doing it all the time, swallowing your gum is really no big deal. It will most likely pass right on through to the toilet bowl.

Myth: Eating turkey makes you sleepy

While not everyone stoops to the level of *Seinfeld*'s Jerry and George, who used the tryptophan in turkey to send a girl to sleep so that they could play with her toys, the supposed sleep-inducing effects of tryptophan in turkey are common currency in America, repeatedly recounted at Thanksgiving dinners and in the popular media around the holidays.

Turkey does contain tryptophan, an essential amino acid. And scientific evidence does support a connection between tryptophan and sleep. L-tryptophan has been marketed as a dietary supplement to ease insomnia. Tryptophan also may have an effect on the immune system, with possible benefits for autoimmune disorders such as multiple sclerosis.

But the truth is turkey is not the cause of your sleepiness. Chicken and minced beef contain almost the same amount of tryptophan as turkey – about 350 milligrams per 100 gram serving. While you might have heard someone claim that turkey made them drowsy, you have probably never heard anyone say that chicken, minced beef or any other meat made them sleepy. Swiss cheese and pork actually contain more tryptophan per gram than turkey, and yet that American classic, a ham and cheese sandwich, somehow escapes blame.

The amount of tryptophan in a single serving of turkey (350 milligrams) is also lower than the amount typically used to induce sleep. The tryptophan supplements to help you sleep contain 500 to 1,000 milligrams. Many scientists also think the limited amount of tryptophan in turkey would be offset by the fact that it is generally eaten in combination with other foods and not on an empty stomach. One clinical trial found comparable results for tryptophan from a food-protein source and pharmaceutical-grade tryptophan, but this study used an extremely rich source of tryptophan, de-oiled gourd seeds, which have twice the tryptophan content of turkey. And in this trial, as in the general use of supplements, tryptophan was taken on an empty stomach to aid absorption. Although we did not locate any experimental evidence to support this claim, many believe that the presence of other proteins and food in the stomach during the feasts generally associated with turkey consumption would limit the absorption of the tryptophan in the turkey.

There are other elements of Christmas and Thanksgiving dinners that can induce drowsiness. Large meals have been shown to cause sleepiness regardless of what is eaten, because the body increases blood flow to the stomach and decreases blood flow to and oxygenation in the brain. Meals that are high in proteins or in carbohydrates may particularly cause drowsiness. And don't forget about the booze. One or two glasses of wine, especially for people who only drink alcohol occasionally, can increase drowsiness.

Myth: Milk makes you phlegmy

In the twelfth century, the physician Moses Maimonides recommended removing milk and dairy products from the diet to help those with breathing or congestion problems. Traditional Chinese medicine also attributes increased 'humidity' or increased mucus production to eating too much dairy produce. Particularly for people suffering from asthma or breathing problems, the idea of eating mucus-producing foods may seem particularly dangerous.

Were these early physicians on track? Is avoiding milk key to avoiding problems with congestion or asthma or even just that phlegmy feeling? Scientists have put the question of dairy foods and mucus production to the test and found no evidence to suggest that milk or dairy products really increases mucus production.

One group of scientists took 125 people, randomly gave them cow's milk or a soya drink, and measured their beliefs about the 'milk mucus' effect and their symptoms. All of the drinks were flavoured with cocoa and peppermint to make it impossible to tell who had the cow's milk and who had the soya drink. Even the people conducting the study did not know who had which drink. In the end, there was no significant difference in problems like a 'coating over the mouth', 'swallowing a lot' or having 'thicker saliva' between those who had the milk and those who had the soya. Interestingly enough, the people who believed in advance that there was a 'milk mucus' effect reported more symptoms with both the soya beverage and the cow's milk.

In another study, researchers took sixty adult volunteers, tried to make them ill with the common cold virus (once again, we're amazed people agreed to do this) and then recorded their daily breathing symptoms and their milk intake over the next ten days. Of the fifty-one people who did get a cold, the weight of the nasal secretions did not increase in those who drank more milk. In fact, drinking milk was not associated with cough, nose symptoms or congestion, even in those with the cold virus.

One group of scientists did show that drinking milk or water increased how much spit or saliva you produce. However, the chemicals that make your saliva thicker were not increased. Furthermore, there were no differences in the mucus content of the saliva after drinking milk than after drinking water.

Still, many people (doctors included!) believe that drinking milk increases phlegm production. In a study of 330 parents, 58.5 per cent believed that drinking milk increases mucus production, and almost a third of them had been told this by their doctor. Seventy people who believed that milk caused increased mucus production reported experiencing clearing of the throat, coughing, swallowing, spitting and post-nasal drip because of drinking milk. However, in all of the aforementioned studies that we reviewed, there was no evidence that milk actually increases mucus production. One possible explanation for the increase in phlegm that some people think they experience after drinking milk is that the milk may lead to 'droplet flocculation' or spreading out of small milk droplets in the saliva. That sensation might be mistaken for mucus − but it's nothing more than milk droplets mixing with saliva.

Myth: Eating bananas attracts mosquitoes, while eating garlic repels them

When a mosquito bites Rachel, the bitten part swells dramatically. And mosquitoes around the world love her. She has had her eye swell shut from a mosquito bite in Kenya and her lip develop a cherry-sized protuberance after a bite in Mexico City. Is there something she is eating that attracts the mosquitoes? Experts disagree about what might attract mosquitoes to Rachel; Aaron has his own theory. Your body releases hundreds of compounds, including over 100 from your breath, and we know almost nothing about most of them. We do know that carbon dioxide and lactic acid in the breath and heat are powerful attractants to mosquitoes. Some perfumes and lotions may also attract them. But bananas? No. As for repelling mosquitoes, nothing eaten or swallowed has ever been proven to repel mosquitoes. The idea that bananas or garlic can affect your chances with the mosquitoes is a myth. After decades of work, the best repellent is still DEET. When used correctly and at safe concentrations of less than 30 per cent, it can keep mosquitoes away for five hours or more. Furthermore, it is safe to use DEET repellents on kids, even on infants over two months. In buggy areas, Rachel does not go outside without it.

Myth: At a picnic, you should avoid foods with mayonnaise

If you long to sample potato salad, devilled eggs or other mayonnaise-laden foods at summer picnics but are afraid of the gastrointestinal woes that might result, you can cast your trepidation to the wind. In 2000, a very comprehensive review was published in the *Journal of Food Protection* that looked for all the reports and studies of illness and death caused by bacteria in store-bought mayonnaise. And they found no reports! The need to avoid mayonnaise is completely unfounded! Mayo gets a bad rap because people used to make it at home using raw eggs, and those raw eggs can contain nasty bacteria called *Salmonella*. However, commercial mayonnaise is much safer; in fact, it's pasteurized and sterile. At a picnic, you are much more likely to get ill from unwashed fruits or vegetables or undercooked meat. In general, you also should avoid foods that have been outside for more than two hours, or less if it's really hot.

Myth: Eating grapefruit burns calories

Wouldn't it be lovely if there were foods you could eat that would actually help you burn calories? Many people believe that grapefruit is one of them. In fact, some diet experts not only endorse eating fresh grapefruit, they also recommend 'grapefruit pills', which usually contain a grapefruit constituent called naringin.

The diet-boosting properties of grapefruit are just a half-truth. Grapefruit may be tied in some way to better weight loss, but scientists are not sure how. We found one scientific study from the Scripps Clinic in California in which obese patients who consumed half a fresh grapefruit before meals, three times a day, lost more weight than those who took placebo pills and apple juice, grapefruit pills and apple juice, or placebo pills and grapefruit juice. The increased weight loss was in the range of one to two pounds or more over twelve weeks. The obese patients who had metabolic syndrome (their bodies are some-what resistant to the effects of insulin) also seemed to lose more weight with the grapefruit juice or grapefruit pills than if they had a placebo pill.

The investigators running the study concluded that grapefruit should be part of a weight-reduction diet, even though they did not know why it was having these effects. The patients who ate grapefruit may have changed their behaviour or altered their diets in ways that the study did not measure.

Other studies and expert panels have concluded that grape-fruit does not burn fat in any unusual way. The American Dietetic Association states that 'this long-held myth is just wishful think-ing. Digestion of any food uses a small amount of energy. But no food – or food component – can 'burn up' the calories in food or 'melt away' body fat. If you lose weight when you add grapefruit to your eating plan, it's probably because you're substituting it for another food that has more calories.'

Another theory supporting the use of grapefruit in the diet is the 'thermic' effect. If a food has a thermic effect, it increases how much energy the body burns up after that food is eaten, or how many calories are needed to digest the food eaten. The size of the meal, what the meal consists of, what you have been eating before, your physical activity, your age and how your body reacts to insulin all influence the thermic effect of food. But in spite of some people's claims, there is no evidence that grapefruit has a greater ability to change this thermic effect than other foods do. In careful, randomized studies of body chemistries, naringin, thought to be the key component of grapefruit as a weight-loss supplement, is shown not to change energy expenditure. Furthermore, a study of 352 overweight men and women found that the 'grapefruit pill' was ineffective in helping people slim down.

If you do decide that you want to start eating more grapefruit, despite the mixed evidence supporting grapefruit as a magic weight-loss tool, you may need to think carefully about any other medications you are taking. Grapefruit has been shown to interact with common drugs, hindering their breakdown by the body and leading to toxicity or bad side effects. Statins for high cholesterol, medicines that prevent heart arrhythmias, medicines that suppress the immune system and calcium-channel blockers can all interact with grapefruit, leading to potentially harmful side effects, so ask your doctor before you add grapefruit to your diet.

Myth: Eating at night makes you fat

Health magazines are full of ways to change your eating habits to help you lose weight. One of the big suggestions is to not eat at night. The rationale seems to be that if you eat later in the evening or before you go to bed, you won't have time to burn off those calories before you go to sleep. Or, perhaps, your metabolism slows down in the evening.

A Swedish study compared eighty-three obese women to ninety-four non-obese women, and found that the obese women ate more meals, and more of their meals were eaten in the afternoon, evening or night. But just because obesity and an increased number of meals at night may be tied together somehow, one does not necessarily cause the other. What it all comes down to is this: people gain weight when they ingest more calories than they burn. The key problem in this study was that these obese women were eating more meals (and more calories) – the issue wasn't the time of day that they ate.

In another study, Swedish men didn't show any evidence of weight gain when they ate at night. In a study of eighty-six obese men and sixty-one average weight men, there were no differences in the timing of when they ate. Another study of sixteen obese subjects investigated whether the timing of when they ate would alter the circadian pattern of energy expenditure (the body's normal pattern of when it burns more calories or less calories). The timing of meals did not change the circadian rhythm of when their energy was used up. In a large study of over 2,500 patients of all weights, the time of day that they ate their meals had nothing to do with whether or not they gained weight.

When all is said and done, eating more than three meals per day has been shown to play a role in obesity. Additionally, some studies have connected skipping breakfast with weight gain. Not only is eating breakfast every day associated with maintaining a healthy body weight, but records of calorie intake also suggest that those who eat breakfast each day do a better job of evenly distributing how much they eat throughout the rest of the day. When you eat three regular meals, you are less likely to overeat at any one particular meal.

The key to weight loss or to maintaining a healthy weight is very simple, and yet so hard for most of us to execute: eat fewer calories than your body burns. Trust us. As long as you do that, it doesn't matter what time of day you eat – you'll lose weight.

Myth: You should drink at least eight glasses of water a day

Everyone from *The Times* to teen magazines to your own doctor have told you that one of the most important things you can do to keep healthy is to make sure you drink at least eight glasses of water a day. We lug around huge water bottles and do our best to force down the recommended eight glasses, often believing that drinking enough water will help us lose weight. The problem with all of this is that there's no medical reason for the recommendation. Some might even call it eight glasses of B.S.

You've heard this myth everywhere, but who first recommended it? The earliest record was back in 1945, when the Food and Nutrition Board of the National Research Council (US) stated that adults should take in about 2.5 litres of water a day, most of this contained in prepared foods. If you ignore the last part of that statement, you would interpret this statement as a mandate to drink eight glasses of water per day (2.5 litres is just over 5 pints, which is close to eight glasses of water). However, even in 1945, the Research Council clearly stated that most of the water you need is already present in the foods you eat.

A few years later, a nutritionist named Dr Frederick Stare also seemingly endorsed drinking eight glasses of water a day, but he said that it could be in the form of coffee, tea, milk, soft drinks or even beer. Plus he also made the point that fruits and vegetables are good sources of water. Even these experts never claimed that you had to get hydrated with plain old water.

There's nothing wrong with liking water, but there is no scientific proof stating that you need to drink anywhere near eight glasses a day. One doctor who has made this his research focus, Dr Heinz Valtin, searched through many electronic databases and also consulted with nutritionists and colleagues who specialize in water balance in the body. In all of his research, and in all of the research we conducted to double-check his work, no scientific evidence could be found to suggest that you need to drink eight glasses of water a day. In fact, scientific studies suggest that you already get enough liquid from what you're drinking and eating on a daily basis. We are not all walking around in a state of dehydration. Extensive US Department of Agriculture surveys measuring how much food and drink more than 15,000 people in fifty states consumed in a period of over three years show that the average person in the U.S. took in 2,188 millilitres (or 4.6 pints) of water a day.

Actually, people should be careful not to drink too *much* water. As demonstrated by a recent and much-publicized death of a young woman who participated in a water-drinking contest hosted by her local radio station, drinking too much can result in water intoxication and even death. Too much water dilutes the normal level of sodium in the blood, causing a condition called hyponatremia, in which brain cells can swell and die. Although not a common problem, it occurs more easily in young infants, which is why paediatricians do not recommend giving babies water (especially when they have diarrhoea).

Myth: If you are thirsty, you are already dehydrated

One reason that many people advocate drinking six to eight glasses of water a day is because they fear that thirst kicks in too late. They are afraid that a person is already dehydrated by the time they start to feel thirsty. If you believe that you don't start feeling thirsty until it's too late, then it makes sense to want to ensure you are getting enough water whether you are thirsty or not.

This is just as much of a myth as the idea that you should drink six to eight glasses of pure water a day. In our review of the medical literature and in reviews conducted by experts in physiology and water balance, we could find no scientific studies claiming that thirst begins after a person is already dehydrated. In contrast, many studies conclude that thirst sets in when the concentration of the body's fluids (technically known as the plasma osmolality) rises a very small amount. For you to feel thirsty, the concentration of the body's fluids does not need to rise into the range that defines dehydration. Even a rise of less than 2 per cent in this concentration can lead to thirst, whereas most experts define dehydration as a rise in plasma osmolality of at least 5 per cent. Thirst sets in when the concentration is still within the normal range.

Your body is incredibly sensitive to changes in how much water it has and how much it needs. Your body releases a hormone called vasopressin when it needs to hold in more water or prevent your kidneys from releasing water (in other words, when it doesn't want you to pee quite as much). The body releases vasopressin in response to even smaller changes in the concentration of your body fluids than the changes that make you feel thirsty. This results in instant, moment-to-moment regulation of your water balance and how much urine you pee out.

Trust your body! A healthy body is wonderful at regulating how much liquid you need, how much vasopressin it releases and when you feel thirsty. A healthy body responds quickly and accurately to the need for more fluids, and so you don't have to force down fluids when you are not thirsty and you don't anticipate any extra fluid needs.

Myth: Caffeinated beverages are dehydrating

One of the reasons that many people advocate drinking lots of water and think that they would not otherwise get enough fluids is that they fear that other beverages 'don't count' towards your fluid requirements. Fizzy drinks, coffee and tea are not supposed to count towards the liquid your body needs because they contain caffeine, and since caffeine is a diuretic, these beverages must dehydrate you.

Diuretics are substances that promote the formation of urine within the kidneys and thus lead to more peeing. Water, cranberry juice and alcohol are weak diuretics. Caffeine is thought to be a diuretic, but the idea that caffeinated drinks have a strong diuretic effect is not well-supported by science. A study back in 1928 showed that caffeine acted as a diuretic in patients at rest, but this landmark study, upon which many doctors and dieticians have based their recommendations, involved only three subjects watched over a period of a few hours. Since 1928, more and more studies have begun to suggest that caffeine is only a mild diuretic that poses no harm to your health or even to how well you perform during exercise. In a series of experiments by A. C. Grandjean and colleagues in eighteen healthy adults, caffeinated beverages did not have any effect on changes in body weight, changes in urine or blood concentration, and changes in the concentration of electrolytes in the body. These scientists concluded that 'advising people to disregard caffeinated beverages as part of the daily fluid intake is not substantiated by the results' of this study. Three review articles have been published that summarize all of the studies looking at the diuretic effects of caffeinated beverages, and all three conclude that caffeinated

beverages and water affect the body water balance similarly during exercise.

Did you get that? Even during exercise, when your body most
needs fluids, caffeinated drinks work just as well as water. In
twenty-three experiments testing caffeine doses and whether
they made people pee more, no difference in urinating was
reported in seventeen of the studies, which included consumption of up to 553 milligrams of caffeine (that's around five cups
of coffee).

Most of the studies looked only at how much peeing you did
and at other measures of body fluid concentration over the
course of a few hours. But others did look at caffeine consumption over a long period. Human water balance is always going to
be challenged by periods of diuresis (or things prompting you to
make more urine) over the course of a given day. But what if you
drink caffeine every day? Will you be chronically dehydrated?
The answer, again, is no (if you are healthy). Only two studies in
the reviews looked at the chronic effects of drinking caffeinated
beverages; they, too, found no indication of dehydration by any
measure used to look at body chemistry and hydration status.

Caffeinated beverages might make you pee a bit more, and
certain fizzy drinks may contain more sugar than you would
want to take in, but they really are a reasonable source of the
fluid your body needs. Remember, your body is very good at
keeping the fluid it needs. Even if caffeinated drinks are not your
favourite choice for other reasons, you should not avoid them
because you think they will dehydrate you.

Myth: You can cure a hangover with …

From aspirin and bananas to Marmite and water, you may hear endless remedies to prevent or treat hangovers from your friends, family members and the Internet. Unfortunately, no scientific evidence supports any cure or effective prevention for alcohol hangovers. In a systematic review of randomized trials asking if anything worked to prevent or treat hangovers, no effective interventions were found in either the traditional or complementary medical literature. While a few very small studies showed minor improvements, these studies did not use a standardized system to measure how people felt. The conclusion of the exhaustive review of all the studies was that nothing that had been studied worked: not propranolol, tropisetron, tolfenamic acid, fructose, glucose, nor dietary supplements including borage, artichoke, prickly pear or a yeast-based product. The effective hangover cure is, unfortunately, a myth. While more recent studies in rats show some potential for new products to alter the mechanisms in your body that might cause hangover symptoms, humans also can get ill from some kinds of 'hangover cures'. A hangover is caused when you drink too much alcohol. Thus the most effective way to avoid a hangover is to drink alcohol only in moderation or not at all.

Myth: Beer before liquor, never been sicker

You may have heard this strategy to avoid hangovers: 'Beer before liquor, never been sicker. Liquor before beer, you're in the clear.' Many people believe that they will feel worse if they drink beer before other types of alcohol. We could not identify any medical or scientific evidence to support this claim. Experts conclude that it is the amount of alcohol and the rate of consumption, not the order of consumption, that affect your level of intoxication and your potential hangover. Furthermore, a larger volume of liquid usually takes longer to consume, and beer generally has less alcohol per litre than spirits. If you are drinking beer first, the increased time it will take you to drink the larger volume of beer should mean that you get intoxicated more slowly than if you were drinking spirits first.

Myth: If you pick up food within five seconds of it hitting the floor, it's safe to eat

First of all, you should know that neither of us is overly paranoid about germs. If you could see the tea mugs we drink out of, you would be amazed that we remain reasonably healthy. In general, it seems harmless enough to eat something that you dropped on the floor for just a few seconds. (Okay, so maybe you wouldn't follow the five-second rule in a public toilet, but other than that it doesn't sound like such a bad idea.)

That is, until you start reading the work of scientists who specialize in food science and microbiology, the studies of how bugs get into our food and stay there. We are not sure how these scientists manage to eat anything without being paranoid, and, unfortunately for us, they decided to put the five-second rule to the test.

Food scientists conducted three experiments to find out what happens when the five-second rule comes up against *Salmonella typhimurium*, a fairly common but nasty bacterium that can cause wretched diarrhoea and vomiting. They tested how well *Salmonella* survived on wood, tile and carpet, and they tested how well it transferred from these surfaces to either bologna sausage or bread. They found that bacteria were still alive after four weeks on dry wood, tile or carpet, and that enough of the bacteria survived to be able to transfer to food. Next, the scientists tested how much time it took for the bacteria to transfer from these different floor surfaces to the food. The worst offender of the five-second rule was bologna on tile. Over 99 per cent of the bacterial cells transferred from the tile to the bologna after just five seconds of the bologna hitting the floor! Transfer from wood was a bit slower (5 to 68 per cent of the bacteria were

transferred) and transfer from carpet was actually not very successful. After hitting the carpet, less than 0.5 per cent of the bacteria transferred to the bologna. When they did transfer, bacteria
moved to the food almost immediately upon contact. By five
seconds, it was too late. Other bacteria, like *Campylobacter* and
Salmonella enteritis, can also survive well on formica, tile, stain-
less steel, wood and cotton cloths, so if you listen to the microbiologists, you can never be too careful about cleaning things up
in the kitchen, and you also can't trust the five-second rule. Bacteria that can make you ill can survive on the floor or other
surfaces for a long time, and they can contaminate other foods
that touch them for only a few seconds.

Bacteria aren't the only thing that could make you ill when
your food hits the floor. A study of pesticides on household surfaces shows that these toxic chemicals can also transfer to foods
like apples, salami and cheese. The pesticides seem to take a little
longer than bacteria, though. The average pesticide was only 1
per cent efficient in transferring over to the food at the one-
minute mark, but up to 83 per cent transferred if it was left on
the floor for sixty minutes. Applying more force to the food (like
throwing it against the floor) also resulted in more pesticide getting on the food, up to 70 per cent at ten minutes on hardwood
flooring when bologna was squished with a force equivalent to
1.5 kg.

Myth: You can chew gum instead of brushing your teeth

The ancient Egyptians, the Aztecs and the early Native Americans all used to chew on tree resin, an early version of today's chewing gum. 'Chicle' was enjoyed for its flavour and may have also been used for teeth-cleaning. The Egyptians also chewed on natron pellets in order to freshen the breath. Listening to advertisements on television about how chewing gum can help to prevent tooth decay and can be used instead of or alongside brushing your teeth, you would think that these ancient populations were on to something with the gum-chewing. Certainly they discovered something pleasurable, but is chewing gum really an effective way to clean your teeth?

Modern science says no. A review of the facts and fiction of gum-chewing found several studies that investigated how well gum worked to get rid of plaque and cut down on the food stuck between your teeth. In a study from the 1960s that examined plaque scores for ten volunteers who were either chewing gum only, using a toothbrush only, using a toothbrush and floss, or using a toothbrush and water, the gum-chewing group did the worst. Gum-chewing was found to be the least effective way to remove plaque. Several subsequent studies from the 1970s and 1980s also investigated whether different kinds of sugar-free gum did a better job of removing plaque. Once again, the answer was no. In some of the studies, plaque was reduced by gum-chewing, but it was not reduced in areas where cavities form. Moreover, the dentists reviewing these studies said that the removed plaque might have a cosmetic effect but not a therapeutic effect (it would have no effect on how many fillings you need). Studies on gum with a special 'abrasive agent' to give the gum extra rough cleaning power showed the same result.

Your mother was right after all – you need to brush your teeth. Chewing gum does not remove plaque well enough to substitute for brushing your teeth, and there is no evidence that it is particularly helpful in addition to brushing.

Myth: You should wait an hour after eating before you go swimming

'You can't go swimming yet!' As a kid who spent most of every summer in the water, Rachel heard this one a lot. The rule was that you had to wait at least one hour before you could jump back in the pool or the lake after you had eaten. We suspect this regulation sprang from the grown-ups' fatigue with keeping an eye on us in the water all the time. But there also seemed to be a genuine fear that if you took so much as a leisurely lap in the pool while food was in your tummy, you might be gripped with horrible cramps that would lead to you drowning right then and there.

Kids will love us and adults may curse us, but there really does not seem to be any proof to back up this claim. In our research, we cannot find any cases of drownings, or near drownings, attributed to eating. While that doesn't mean it could never happen, there is no proof that this is a real danger. Expert groups don't really say that you have to wait to go swimming either. Neither the American Academy of Pediatrics nor the American Red Cross has any recommendations about how long you should wait after eating to swim. As early as 1961, expert exercise physiologists were already saying that this idea of getting cramps while swimming after eating was 'questionable'.

Isn't it still possible? Couldn't you get horrible cramps because your body is so busy trying to digest the food in your stomach? It is true that the digestive system diverts some of the circulating blood away from your muscles and towards your gut when you have eaten and need to digest food. As with any exercise immediately after eating, you might feel somewhat uncomfortable after a big meal. But, even if you were to get cramps, it is very unlikely that you would be completely incapacitated. If you get cramps, you should just get out of the water and give your body a rest. Whether you have been wolfing down sandwiches or not, you should not be swimming in dangerous places from which you have no ability to escape if your body is tired or you have muscle cramps. Making sure that you or your children only swim in safe places should give you peace of mind, even if you end up getting cramps.

Myth: It's safe to double-dip

Although eating from communal bowls has probably been around for ever, the first reference to 'double-dipping' seems to have come from a 1993 *Seinfeld* episode in which the legendary George Constanza is berated after he dips the same chip twice at a gathering after a funeral. His girlfriend's brother, Timmy, asks, 'Did you just double-dip that chip?' Timmy claims that the offence of double-dipping is 'like putting your whole mouth right in the dip'. George doesn't think it's such a big deal. (Of course, he also eats out of the dustbin in another episode.) 'You dip the way you want to dip, I'll dip the way I want to dip,' he snapped. And if you know George, you know what happened next – they end up wrestling over the contaminated chip. At the funeral.

Salsa. Guacamole. Hummus. Whether you dunked in a spicy or sweet or creamy dip, whether you used a crisp or a cracker or a vegetable, we bet that you have been a double-dipper too. You wanted just a bit more of something on the rest of your crisp, and so you plunged it back into the dip a second time. Sticking a crisp back into the bowl of dip doesn't seem like a big thing, right? You might argue that we are exposed to small amounts of germs all the time.

Once again, the food scientists want to ruin our fun – and it seems they side with Timmy on this one. One intrepid group of microbiologists, led by Dr Paul Dawson, studied whether or not bacteria were really transferred from mouth to chip to dip (the double-dipping scenario). Volunteers took a bite of a wheat cracker and then dipped the cracker for three seconds into a tablespoon of a test dip. They used six different dips: salsa, cheese dip, chocolate syrup, and sterile water with three different acidity levels. The scientists compared the double-dips to what happened when a brand-new cracker was dipped, and they measured the amount of bacteria in the volunteers' mouths. And the results were not pretty.

On average, three to six double-dips transferred about 10,000 bacteria from the eater's mouth to the dip. And each cracker picked up between one and two grams of dip. If you are at a party and three to six people double-dip their chips, any chip you dip may pick up at least 50 to 100 bacteria. Could this make you ill? Maybe. It depends on how many bacteria you pick up in your bite, how many people are double-dipping and what bacteria they have in their mouths. This study only looked for some types of bacteria, and other kinds of bacteria might have been in the dip. The risk to you might depend on whether the bacteria in the dip are the kind that can make you really ill or a less harmful variety. However, the study revealed that bacteria can be transferred and that the dip does not kill off all of the bacteria.

For the germ-phobes and paranoid folk out there, it may be helpful to know that some kinds of dips are less likely to harbour bacteria. The more acidic water samples had less bacteria and the number of bacteria in them got smaller over time. But acidity is not everything. An acidic salsa sample picked up more bacteria than the cheese or chocolate dips. This is probably because it was runny. The thicker the dip, the more of it sticks to the chip, which can act to seal in the bacteria. So fewer bacteria are left behind in the bowl. Thus thicker dips may be safer after double-dipping than thinner dips.

The next time you are at a party and considering a plunge into the dip bowl, Dr Dawson suggests you take a look at the people around you. Would you kiss them? Would you ever want to lick the insides of their mouths? If you don't want to swap spit with these people, then think twice before helping yourself to the dip.

PART VI

'Vaccines made my baby autistic'

Myths That Spark Controversy and Debate

Myth: The fluoride in your water is dangerous

Sixty years ago, Rachel's hometown, Grand Rapids, Michigan, became the first city in the world to implement fluoridation of the entire water supply, which means that they added more fluoride to the already naturally occurring fluoride in their water source. Grand Rapids paved the way, but thousands of other communities have since added fluoride to their water supplies, with the aim of raising the fluoride level to the optimum amount that will prevent cavities and tooth decay. Two-thirds of the US population lives in a community where the water is fluoridated. However, some believe that this is a cause for concern, not celebration.

The Fluoride Action Network, an organization dedicated to educating people about the dangers of fluoride, offers many arguments against the policy of adding fluoride to water supplies. Among their claims, they note that fluoride has minimal benefit when it is swallowed and that fluoride is not recommended for babies. They propose that fluoride has many risks, including to the brain, thyroid gland and bones. Others make much more alarmist statements, calling fluoride a 'corrosive poison'. Websites and publications raising concerns about fluoride offer 'scientific references' and 'expert quotes' that really could cause concern that the fluoride in your water not only doesn't help you but could actually be hurting you. Let's review the evidence.

First of all, overwhelming evidence has existed for over sixty years that proves the efficacy of water fluoridation. Some of the best studies come from the early days of fluoridation, because it was easier then to find groups of people who were not exposed to any fluoride sources. In a fifteen-year landmark study from Grand Rapids, Michigan, children who received fluoridated water from birth had 50–63 per cent less tooth decay than children from a nearby city in Michigan who drank non-fluoridated water. A comparison study in Newburgh, New York, found that six- to nine-year-olds had 58 per cent less tooth decay than kids in a nearby city with fluoride-deficient water ten years later, and they had 70 per cent less decay fifteen years later. A huge survey in the United States of almost 40,000 school children found that tooth-decay rates were declining overall (probably because of higher-quality toothpastes, mouthwashes and supplements that all contain fluoride), but even when scientists controlled for all of those other things the kids might be exposed to, kids who had fluoride in their drinking water still had 25 per cent less tooth decay.

The evidence for how well fluoride works is overwhelming. A compilation of the results of 113 studies in 23 countries showed reductions in tooth decay for both baby teeth and adult teeth. People with fluoride in their water had 40–49 per cent less tooth decay in baby teeth and 50–59 per cent less tooth decay in adult teeth. Another compilation of studies conducted between 1976 and 1987 showed reductions in tooth decay of between 15 and 60 per cent, with the highest benefit for baby teeth, but also significant benefits for permanent teeth. Yet another systematic review of the literature analysed 214 studies to determine whether fluoride in drinking water was effective. This systematic review found that when you combine all 214 of those studies, you discover that children who got fluoride in their drinking water had fewer teeth with cavities, and were more likely to have no cavities at all, compared to the children who did not have fluoride in their drinking water. The 350 peer-reviewed references compiled by the American Dental Association in their publication *Fluoride Facts* also support how well fluoride works.

Of course, preventing cavities might not be worth endangering your health in other ways. Some people are concerned about water fluoridation leading to cancer, thyroid problems, neurological problems and heart disease, and other toxicities. Is fluoride really safe? Once again, the scientific evidence overwhelmingly supports the safety of adding fluoride to a community water supply. Reviews of the safety of fluoride by the Institute of Medicine's Food and Nutrition Board, the National Research Council, the Department of Health and Human Services and the Public Health Service in the United States, the British National Health Service and the World Health Organization have all led to the conclusion that fluoride supplementation is safe, effective and recommended for community water supplies. A systematic review published in the *British Medical Journal* in 2000 analysed 214 studies on fluoridation and found no evidence of potential adverse effects except for dental fluorosis (which we'll talk about in a moment). Periodic reviews every six years by the US Environmental Protection Agency continue to find no harmful effects related to fluoride in drinking water. Good scientific studies demonstrate that drinking fluoridated drinking water does not increase the risk of hip fractures. More than fifty extremely large studies do not show any association between fluoridation and the risk of cancer. One small study from the 1950s of fifteen patients with an overly active thyroid (hyperthyroidism) tried to use large amounts of fluoride as a treatment for hyperthyroidism and found that it seemed to help some patients. On that basis, concerns have been raised about whether fluoride in drinking water adversely affects the thyroid gland. Again, much better science says that the answer is no. Studies of people with drinking water with naturally high levels of fluoride found that it had no effect on the size or function of their thyroid gland, and this matches results from animal studies. Furthermore, two studies have found no association between the level of fluoride in water and thyroid cancer.

The National Research Council of the US National Academy of Sciences supports the conclusion that drinking optimally fluoridated water is not a genetic hazard. There is no known association between drinking fluoridated water and Down's Syndrome. One psychiatrist in the 1950s published two studies claiming that the two were connected, but four subsequent studies have found no connection and experienced researchers have noted significant problems with how that psychiatrist analysed his data. There is also no generally accepted scientific evidence establishing a link between fluoridated water and other neurological disorders, including attention-deficit disorder.

One study, in which rats were administered fluoride at 125 times the level in community fluoridated water, concluded that the rats showed some behavioural changes. However, this study did not use any sort of control group to compare the rats that had the fluoride to rats that did not, and scientists who reviewed the results of this study have concluded that it is significantly flawed and cannot be used to draw conclusions about problems with fluoridated water. Moreover, a seven-year-long study in human children from birth to age six found no effects from the fluoridated water on the children's health or behaviour, using both mothers' ratings and teachers' ratings of the children's behaviour.

There is one real problem that can result from too much fluoride – dental fluorosis. This is a discolouration of the teeth that can occur when a child ingests more fluoride than is recommended. With mild dental fluorosis, the teeth get white flecks or spots, but with severe fluorosis the teeth can get a permanent brown stain. About 10 per cent of the mild fluorosis seen in children does probably come from the fluoridation of water (although the dentists argue that these small white flecks are a small price to pay for avoiding cavities, tooth decay, missed school and so on). The biggest problem with fluorosis is that kids do sometimes have a habit of swallowing their toothpaste. Fluorosis is the reason that the American Dental Association recommends that children under six do not use more than a pea-sized amount of toothpaste. We assume that little kids are going to swallow their toothpaste, and if they swallow a lot more than the pea-sized amount, they are more likely to get fluorosis. Toothpaste delivers a much more concentrated amount of fluoride than the drinking water does. The biggest cause of fluorosis is most likely swallowed toothpaste, not the water.

Many experts have concluded that fluoride is both beneficial and safe in the drinking water of our communities, and they agree that the risk of fluorosis is far outweighed by the benefits of preventing tooth decay. The American Dental Association, the US Centers for Disease Control and Prevention, the American Medical Association and the National Health Service have all issued statements supporting how well adding fluoride to water works to prevent tooth decay. In fact, the Centers for Disease Control and Prevention declared fluoridation of public drinking water to be one of the ten biggest public health achievements of the twentieth century. We agree. Grand Rapids had the right idea!

Myth: It is safe for babies to sleep in bed with their parents

There are some myths that we really didn't want to investigate because we knew that our friends and family would be upset with us for trying to disprove one of their most cherished beliefs. Here comes one of those beliefs – the safety of having your baby sleep in bed with you (co-sleeping).

Sharing a bed with your baby is much more common in some communities than in others. One study of 128 parent-infant pairs in the Washington, D.C., area found that nearly half (48 per cent) of infants slept in an adult bed with their mother. In some countries, co-sleeping is even more common. Some advocate that sharing a bed promotes breastfeeding. Other studies suggest that sharing a bed may enhance infant arousal sleep patterns that could help prevent Sudden Infant Death Syndrome (SIDS). More breastfeeding and less SIDS? Co-sleeping sounds like a perfect choice – so why don't we recommend co-sleeping to everyone?

The problem is that sharing an adult bed with a baby can be dangerous. The soft bed, pillows and loose blankets on an adult bed can entrap and suffocate babies, especially those who are too young to roll over on their own. The fact is babies have died and continue to die every year from co-sleeping. A review of 1,396 cases of accidental suffocation of infants eleven months old and younger reported to the United States Consumer Product Safety Commission between 1980 and 1998 found that the number of infant suffocations in adult beds, sofas and chairs increased each year. In this study, infants who slept in adult beds had a twenty-fold increased risk of suffocation compared to those who slept in cribs.

The National Institute of Child Health and Human Development has not identified any scientific proof that SIDS is reduced by co-sleeping; instead, they conclude that sharing a bed may be unsafe for babies because of the risk of entrapment and suffocation. The Consumer Product Safety Commission and The American Academy of Pediatrics Task Force on Infant Positioning and SIDS both oppose co-sleeping due to the risk of infant suffocation.

The National Health Service does not warn parents against co-sleeping. They do recommend that you do so only if you are not tired, drugged or drunk. However, UNICEF UK point out that adult beds are not designed to keep a baby safe and they document a number of things that make co-sleeping in an adult bed potentially unsafe.

While having your baby sleep in your bed may make it easier for you to feed them, comfort them and get some rest yourself, these advantages should be weighed against the very real risk of the baby suffocating and dying.

Myth: More people commit suicide around public holidays

Public holidays can bring out the worst in us – even as we gather with our family and friends for the purpose of celebrating. Maybe the holidays just put too much pressure on us to pull things together or to put on a show for others. Maybe our families drive us absolutely crazy in ways that only our loved ones can. Or maybe we don't have friends or families to celebrate with, and the loneliness is much worse around the festivities. Plus many of our public holidays fall in the cold, dark, winter months, when we expect rates of depression to be higher. For whatever the reason, it makes sense to think that more people commit suicide around the holidays than at other times of the year. After all, the holidays can be a very depressing time for those who have lost loved ones or who are disappointed with themselves, in contrast to the holiday cheer. The media has reported on the link between holidays and suicides time and again.

While the popular assumption is that holidays are a risk factor for suicide, people aren't really any more likely to kill themselves around the holidays than at any other time of year. In a study from Japan that looked at suicides between 1979 and 1994, the rate of suicide was the lowest in the days before a holiday, but the highest in the days after the holiday. In contrast, in a study from the United States, the number of suicides within a thirty-five-year period did not increase before, during or after holidays – including birthdays, Thanksgiving, Christmas, New Year's Day or the Fourth of July. However, a smaller study of adolescents in a different part of the United States did show an association between the dates when teenagers attempted suicides and the occurrence of holidays, with a peak in suicide attempts at the end of the school year. Interestingly, in the United States, psychiatric visits actually decrease before Christmas and increase again afterwards. Researchers speculated that this may actually reflect increased emotional and social support during holidays. The US Centers for Disease Control concluded that holidays do not increase the risk for suicide. Suicide data from Ireland from 1990 to 1998 also failed to connect suicides with the holidays. While Irish women were no more likely to commit suicide on holidays than on any other days, Irish men were actually significantly less likely to commit suicide on holidays.

Furthermore, people are not more likely to commit suicide during the heights of winter darkness and doldrums. Around the world, suicides peak in warmer months and are actually at their lowest in the winter. This pattern was reported as early as 1897 by the sociologist Emile Durkheim, who described how European 'suicide reaches its maximum during the fine season, when nature is most smiling and the temperature mildest'. In Finland, a country of long and dark winters, suicides peak in autumn, and are actually at their very lowest in the winter. In a thirty-year study of suicides in Hungary between 1970 and 2000, researchers again found peaks of suicide in the summer and the lowest rate of suicides in the winter. Studies of suicide rates from India also reveal peaks in April and May. Studies from the United States reflect this same pattern, with lower rates in November and December than in typically warmer months.

While we try to debunk myths wherever we can find them, we do not want to leave you convinced that suicides won't happen at particular times of the year. Suicides, and the depressions that often lead up to them, are devastating events at any time. Regardless of whether or not it's a holiday, the middle of winter or bright and sunny summer, thoughts of wanting to put an end to things or to kill oneself must be taken seriously. If you or someone you know is having these thoughts, seek professional attention immediately.

Myth: Newer drugs are always better

As a society, we're addicted to drugs. Almost all of them are legal, and we're not abusing them per se, but we want them desperately. The problem is that so many of the new prescription drugs we take are no better than old drugs that are less expensive. Since new drugs are almost always more expensive, we're wasting money. In some cases, the older drugs are actually better – meaning that we're spending more for less benefit. Even more concerning, 'new' drugs sometimes aren't actually new at all, making their production and marketing suspect at best.

We want to start with a caveat. We don't hate drug companies. We don't hate people who work for drug companies. We don't even hate drugs. In fact, both of us, as practising doctors, have seen drugs save lives, improve health and make daily life incredibly better. But that doesn't mean the pharmaceutical industry gets a free ride.

Often, completely new drugs come to market along with a huge advertising campaign and the promise of research showing their effectiveness. The problem is that to get MHRA approval drug companies only need to show that their drug is more effective than a placebo. That's right – effective doesn't mean better than what is already available, it means better than nothing. And often, unless a drug company pays for a head-to-head comparison, this type of research just won't happen.

Once in a blue moon, however, these studies do happen. The biggest and best of them was the Antihypertensive and Lipid-Lowering Treatment to Prevent Heart Attack Trial (ALLHAT). Drugs for high blood pressure are intended to reduce the risk of complications or death due to coronary-artery or other cardiovascular disease. There were so many drugs to choose from (at different costs) that the National Heart, Lung, and Blood Institute (NHLBI) organized and supported a randomized controlled trial to examine which was best. This study was enormous; it took place in 623 centres in the United States, Canada, Puerto Rico and the US Virgin Islands between 1994 and 1998, and included over 33,000 participants. Patients received one of four drugs:

1. amlodipine, a calcium-channel blocker;
2. doxazosin, an alpha-adrenergic blocker;
3. lisinopril, an angiotensin-converting enzyme inhibitor;
4. chlortalidone, a diuretic.

The last of these, the diuretic, was the oldest of the drugs, and by far the cheapest. However, at the end of the study, the results were clear. This old, cheap diuretic was significantly better at preventing at least one of the major types of cardiovascular disease when compared to the other, newer drugs. Since the diuretic was also significantly less expensive, it should be the drug of choice in initial treatment of high blood pressure. However, it usually is not.

The other drugs were good-faith efforts to create new molecules to treat a chronic disease. However, in many other instances, new drugs are just 'changed' old drugs with no expectation that they will be better. When creating drugs through organic synthesis, mirror-image molecules are created. If drug D is created, you wind up with a compound consisting of half D and half D´ (the mirror image of D). The mirror image is usually inert and has no effect on the drug or the individual taking the drug, but it is left in because there is an expense to remove it. Years ago, the drug companies hit upon a brilliant idea. If they removed that non-working, mirror-image part of the pill, they could claim they devised a new drug!

Think this is rare? Ever heard of Nexium ('the purple pill')?

Nexium is just Losec, with the mirror-image part removed. And Losec is an effective, and now generic, drug for heartburn. Losec is D + D´; Nexium is just D. There is no reason to believe that equivalent amounts of the two drugs are not the same – and research supports this. Four head-to-head studies compared 20 milligrams of Losec to 20 or 40 milligrams of Nexium. But you have to remember – half of Losec is D´(filler)! So these studies really compared 10 milligrams of D to 20 or 40 milligrams of P. Shouldn't more be better? One would think so, but it was barely so, and only in half the studies. And, of course, none of the advertising stated that you could get the same improvement just by taking more Losec.

This isn't the only offender. In fact, since 1990, the proportion of these 'half' drugs among approved new drugs worldwide has become greater than half of those new approvals.

Myth: Vaccines cause autism

We have no desire to downplay the impact of autism on the families and lives of children it affects. We have no desire to dissuade research into the causes of autism and why it appears to be more prevalent in recent years. However, we have to believe what science tells us, even when we dislike the results. Good science is based on widely accepted principles and methods; anecdotal evidence is not research. Believe us – we have no great love of the pharmaceutical industry, nor do we have any vested interest in keeping vaccines going. This book should stand as a testament to our scepticism of accepted wisdom. In this case, however, conventional wisdom appears correct. Science does not support a link between vaccines and autism.

The idea that vaccines cause autism began in 1998, when an article was published in *The Lancet* that followed the cases of twelve children with developmental regression and gastrointestinal symptoms, such as diarrhoea or stomach pain. Nine of those children had autism, and eight of the nine had parents who thought the symptoms of autism developed after the vaccine for measles, mumps and rubella (MMR) was administered. This was not a randomized controlled trial, or even a scientific study. It was merely a description of a small group of children. To be honest, it's difficult to imagine such a study getting published in *The Lancet* today. Based on the described beliefs of those eight parents, a frenzy of fear about vaccines and autism has ensued for the past decade. Moreover, these concerns about autism and vaccines are only heightened by a timing issue. Remember, humans try to make sense of the world by seeing patterns. When they see a disease that tends to appear around the time a child is a year old (as autism does), which is also the age at which children get particular vaccinations, our human brains want to put those things together. But just because two things happen at the same time, one does not cause the other. This is why we need careful, scientific studies to answer important questions like this. There have been many such studies in the last decade that have contradicted the hypothesis that vaccines cause autism. These are just a few:

- In 1999, a study in *The Lancet* described almost 500 children with autism born in England after 1989. No difference was seen in the age of diagnosis for those who did and those who did not receive vaccines. This means that either there is no association between the MMR vaccine and autism or the association is so weak that it could not be detected in a large sample of children with autism.
- A 2001 study published in the *Journal of the American Medical Association* described data on over 10,000 kindergarten children in California from 1980 to 1994. The incidence of autism over that time (calculated from birth rates in California) increased over that time from 44 per 100,000 births to 208 per 100,000 births – a 373 per cent increase. The increase in MMR coverage, however, rose from 72 per cent to 82 per cent, a much smaller increase of 14 per cent. Thus the relatively small increase in children getting the MMR vaccine is too small to be responsible for the very large increase in autism.
- A 2002 study in the *New England Journal of Medicine* followed all children born in Denmark from 1991 to 1998. This means they obtained data on over 530,000 children born in those years. They could find no association between the development of autism and the age at vaccination, the time since vaccination or even the date of vaccination. (Just for the record, that is 530,000 children followed scientifically versus the beliefs of eight parents!)
- In 2005, a systematic review of studies examining the

effectiveness and unintended effects of the MMR vaccine was published in the Cochrane Database. They identified 139 potential studies, and 31 met the criteria for their review. After a thorough investigation of these studies, even though the MMR could be associated with a number of side effects or other issues, they could find no evidence for an association between the vaccine and autism.

Even with overwhelming evidence to the contrary, people still claim that the jury is out about vaccines and autism. The recent case of Hannah Poling has reignited the debate. Hannah is a girl with an underlying mitochondrial enzyme deficiency who developed encephalopathy caused, her parents believe, by vaccinations. This encephalopathy led to long-term symptoms that fall under the autistic spectrum. After a great deal of effort, the Vaccine Injury Compensation Program, a US government program responsible for compensating patients who suffer from complications arising from vaccines, agreed to hear her case. This decision was not based on evidence. There is *still* no good body of scientific evidence to support this belief. However, the fact that the Vaccine Injury Compensation Program even agreed to hear the case was heralded by some as a concession by the government that vaccines may cause autism. To refute this, after a press conference to discuss the case, Julie Gerberding, the director of the Centers for Disease Control and Prevention, said, 'The government has made absolutely no statement indicating that vaccines are a cause of autism [... t]hat is a complete mischaracterization of the findings of the case and a complete mischaracterization of any of the science that we have at our disposal today.'

Decisions such as this continue to fuel the efforts of those who believe that an association exists, even in the face of the scientific evidence. Unfortunately, unlike many of the myths described earlier in this book, believing this myth has potentially serious health consequences. Since this controversy began, many people have decided not to have their children immunized. As a result, more people are getting the diseases that vaccines prevent, sometimes with devastating consequences. Children can get very, very ill from measles, mumps or rubella. Sometimes they even die. Rachel works in Kenya for much of the year, and she routinely sees very ill children who are suffering precisely because they did not get vaccines. These diseases are still around, and as people travel more and more, children and adults in any country can be exposed to them. At least two million people of all age groups die every year from diseases that could have been prevented with existing vaccines.

Remember, this controversy all started ten years ago with a paper describing the beliefs of parents of eight children with autism. Since that time, ten of the twelve authors of that paper have publicly and professionally retracted from the original paper the supposition that MMR could cause autism; this is a rare occurrence in medical literature. An eleventh author could not be contacted before the release of the retraction in 2004. And the final author, who was the lead author of the original study, was investigated earlier this year for ethical violations, allegations of professional misconduct and undisclosed conflicts of interest in conducting that original research. The investigation is ongoing.

Imagine how different the world would have been if that one small study hadn't been published.

References

Men with big feet have bigger penises

1. Cobb, J., and D. Duboule (2005), 'Comparative analysis of genes downstream of the Hoxd cluster in developing digits and external genitalia', *Development* 132 (13), pp. 3055–67.
2. Edwards, R. (1998), 'Definitive Penis Size Survey' (2nd edn), http://www.sizesurvey.com/result.html (accessed 31 March 2008).
3. Furr, K. (1991), 'Penis size and magnitude of erectile change as spurious factors in estimating sexual arousal', *Annals of Sex Research* 4 (3), pp. 265–79.
4. Gebhard, P., and A. Johnson (1979; reprint edn 1996), *The Kinsey Data: Marginal Tabulations of the 1938–1962 Interviews Conducted by the Institute for Sex Research* (Bloomington, Ind.: Indiana University Press).
5. Lee, P. A. (1996), 'Survey report: concept of penis size', *Journal of Sex and Marital Therapy* 22 (2), pp. 131–5.
6. Shah, J., and N. Christopher (2002), 'Can shoe size predict penile length?', *BJU International* 90 (6), pp. 586–7.
7. Siminoski, K., and J. Bain (2004), 'The relationships among height, penile length, and foot size', *Sexual Abuse: A Journal of Research and Treatment* 6 (3), pp. 231–5.
8. Spyropoulos, E., et al. (2002), 'Size of external genital organs and somatometric parameters among physically normal men younger than 40 years old', *Urology* 60 (3), pp. 485–9.

You use only 10 per cent of your brain

1. Baranaga, M. (1997), 'New imaging methods provide a better view into the brain', *Science* 276, pp. 1974–6.

2. Beyerstein, B. L. (1999), 'Whence Cometh the Myth That We Only Use Ten Percent of Our Brains?', in S. Della Sala (ed.), *Mind-Myths: Exploring Popular Assumptions About the Mind and Brain* (Indianapolis, Ind.: John Wiley & Sons).

3. Carnegie, D. (1944), *How to Stop Worrying and Start Living* (New York: Simon & Schuster).

4. Damasio, H., and A. R. Damasio (1989), *Lesion Analysis in Neuropsychology* (New York: Oxford University Press).

5. Gazzaniga, M. (1989), 'Organization of the human brain', *Science* 245, pp. 947–52.

6. Langa, K. M., et al. (2008), 'Trends in the prevalence and mortality of cognitive impairment in the United States: is there evidence of a compression of cognitive morbidity?', *Alzheimer's and Dementia* 4 (2), pp. 134–44.

7. Marden, O. S. (1909), *Peace, Power, and Plenty* (New York: Thomas Y. Cromwell).

8. Marden, O. S. (1917), *How to Get What You Want* (New York: Thomas Y. Cromwell).

9. Marg, E., J. E. Adams and B. Rutkin (1968), 'Receptive fields of cells in the human visual cortex', *Experientia* 24, pp. 348–50.

10. Petersen, S. E., et al. (1990), 'Activation of extrastriate and frontal cortex areas by visual words and word-like stimuli', *Science* 249, pp. 1041–44.

11. Roland, P. E. (1993), *Brain Activation* (New York: Wiley–Liss).

12. Rosner, B. S. (1974), 'Recovery of Function and Localization of Function in Historical Perspective', in D. G. Stein, H. Rosen and N. Butters (eds.), *Plasticity and Recovery of Function in the Central Nervous System* (New York: Academic Press).

13. Sacks, O. (1985), *The Man Who Mistook His Wife for a Hat and Other Clinical Tales* (New York: Summit Books).

Your hair and fingernails continue to grow after you die

1. Christoph, T., et al. (2000), 'The human hair follicle immune system: cellular composition and immune privilege', *British Journal of Dermatology* 142 (5), pp. 862–73.
2. Maple, W., and M. Browning (1994), *Dead Men Do Tell Tales* (New York: Doubleday).
3. Paus, R., and G. Cotsarelis (1999), 'The biology of hair follicles', *New England Journal of Medicine* 341 (7), pp. 491–7.
4. Remarque, E. M. (1929), *All Quiet on the Western Front, trans. A. W. Wheen* (Boston, Mass.: Little, Brown & Co.).
5. Snopes.com, 'Fingernails grow after death', http://www.snopes.com/science/nailgrow.asp (accessed 11 June 2008).

If you shave your hair, it will grow back faster, darker and thicker

1. Lynfield, Y. L., and P. Macwilliams (1970), 'Shaving and hair growth', *Journal of Investigative Dermatology* 55 (3), pp. 170–72.
2. Paus, R., and G. Cotsarelis (1999), 'The biology of hair follicles', *New England Journal of Medicine* 341 (7), pp. 491–7.
3. Saitoh, M., M. Uzuka and M. Sakamoto (1970), 'Human hair cycle', *Journal of Investigative Dermatology* 54 (1), pp. 65–81.
4. Trotter, M. (1928), 'Hair growth and shaving', *Anatomical Record* 37 (Dec.), pp. 373–9.
5. Trueb, R. M. (2002), 'Causes and management of hypertrichosis', *American Journal of Clinical Dermatology* 3 (9), pp. 617–27.

If you pull out a grey hair, two grow back in its place

1. Tobin, D. J., and R. Paus (2001), 'Graying: gerontobiology of the hair follicle pigmentary unit', *Experimental Gerontology* 36 (1), pp. 29–54.

2.   Wolff, K., et al. (eds.) (2008), *Fitzpatrick's Dermatology in General Medicine*, 7th edn (New York: McGraw-Hill Professional).

You'll ruin your eyesight if you read in the dark

1.   Fredrick, D. R. (2002), 'Myopia', *British Medical Journal* 324 (7347), pp. 1195–9.
2.   Goto, E., et al. (2002), 'Impaired functional visual acuity of dry eye patients', *American Journal of Ophthalmology* 133 (2), pp. 181–6.
3.   Howstuffworks.com, 'Does reading in low light hurt your eyes?', http://science.howstuffworks.com/question462.htm (accessed 11 June 2008).
4.   Kirkwood, B. J. (2006), 'Why do humans blink? A short review', *Insight* 31 (3), pp. 15–17.
5.   ninemsn.com, 'Does reading in dim light ruin your eyes?', http://health.ninemsn.com.au/article.aspx?id=113116 (accessed 11 June 2008).
6.   Rubin, M. L., and L. A. Winograd (2003), *Taking Care of Your Eyes: A Collection of Patient Education Handouts Used by America's Leading Eye Doctors* (Gainesville, Fla.: Triad Publishing Company).
7.   Sheedy, J. E., S. Gowrisankaran and J. R. Hayes (2005), 'Blink rate decreases with eyelid squint', *Optometry and Vision Science* 82 (10), pp. 905–11.

If you don't shut your eyes when you sneeze, your eyeball will pop out

1.   *MythBusters*, episode 84 (2007): 'Viewers' Choice Special', http://mythbustersresults.com/episode84 (accessed 20 May 2008).
2.   *New York Times*, 30 April 1882, 'Burst an eyeball in sneezing'.

The average person swallows eight spiders a year

1.   Gilmore, B. (2008), 'Chances of eating a spider while asleep', http://www.brownreclusespider.org/eating-spider-while-asleep. htm (accessed 9 May 2008).
2.   Holst L. B. (1993), 'Reading is believing', *PC Professional*, p. 71.
3.   Spiderzrule, 'Questions and Answers', http://www.spiderzrule. com/answers.htm (accessed 9 May 2008).

You should move your bowels at least once a day

1.   Benninga, M., et al. (2005), 'The Paris Consensus on Childhood Constipation Terminology (PACCT) Group', *Journal of Pediatric Gastroenterology and Nutrition* 40 (3), pp. 273–5.
2.   North American Society for Pediatric Gastroenterology, Hepatology, and Nutrition (2006), 'Evaluation and treatment of constipation in infants and children: recommendations of the North American Society for Pediatric Gastroenterology, Hepatology, and Nutrition', *Journal of Pediatric Gastroenterology and Nutrition* 43 (3), e1–13.
3.   ROME Foundation, 'Diagnostic Criteria for Functional Gastrointestinal Disorders', http://www.romecriteria.org/documents/ Rome_II_App_A.pdf (accessed 9 May 2008).

Your urine should be almost clear

1.   Bulloch, B., et al. (2000), 'Can urine clarity exclude the diagnosis of urinary tract infection?', *Pediatrics* 106 (5), e60.
2.   Valtin, H. (2002), '"Drink at least eight glasses of water a day." Really? Is there scientific evidence for "8 x 8"?', *American Journal of Physiology – Regulatory, Integrative and Comparative Physiology* 283 (5), pp. 993–1004.

You lose most of your body heat through your head

1. O'Connor, A. (26 October 2004), 'The claim: you lose most of your body heat through your head', *New York Times*, http://www.nytimes. com/2004/10/26/health/26real.html (accessed 31 October 2008).
2. Pretorius, T., et al. (2006), 'Thermal effects of whole head submersion in cold water on non-shivering humans', *Journal of Applied Physiology* 101 (2), pp. 669–75.
3. US Department of the Army (1970), *US Army Survival Manual: FM 21–76*, p. 148.

You can beat a breathalyser test

1. Canoe CNEWS (2005), 'Potty-mouth man can't fool science', http://cnews.canoe.ca/CNEWS/WeirdNews/ 2005/03/30/976892-sun.html (accessed 12 May 2008).
2. Discovery Channel, Mythbusters-Wiki, 'Beat the Breath Test', http://mythbusters-wiki.discovery.com/page/ Beat+the+Breath+Test (accessed 12 May 2008).
3. Sutton, L. R. (1989), 'Evidential breath ethanol analyzers, accuracy and sensitivity to breath acetone', *Blutalkohol* 26 (1), pp. 15–27.
4. Worner, T. M., and J. Prabakaran (1985), 'The accuracy of breath alcohol analysis using the breathalyzer', *Alcohol and Alcoholism* 20 (3), pp. 349–50.

You should never wake a sleepwalker

1. Guilleminault, C., et al. (2003), 'Sleepwalking and sleep terrors in prepubertal children: what triggers them?', *Pediatrics* 111 (1), e17–25.
2. Masand, P., A. P. Popli and J. B. Weilburg (1995), 'Sleepwalking', *American Family Physician* 51 (3), pp. 649–54.

3. Ohayon, M. M., C. Guilleminault and R. G. Priest (1999), 'Night terrors, sleepwalking, and confusional arousals in the general population: their frequency and relationship to other sleep and mental disorders', *Journal of Clinical Psychiatry* 60 (4), pp. 268–76.

Cold or wet weather makes you ill

1. Eccles, R. (2002), 'Acute cooling of the body surface and the common cold', *Rhinology* 40 (3), pp. 109–14.
2. Eccles, R. (2002), 'An explanation for the seasonality of acute upper respiratory tract viral infections', *Acta Oto-Laryngologica* 122 (2), pp. 183–91.
3. Johnson, C., and R. Eccles (2005), 'Acute cooling of the feet and the onset of common cold symptoms', *Family Practice* 22 (6), pp. 608–13.

You can get a hernia by lifting something heavy

1. Pathak, S., and G. J. Poston (2006), 'It is highly unlikely that the development of an abdominal wall hernia can be attributable to a single strenuous event', *Annals of The Royal College of Surgeons of England* 88 (2), pp. 168–71.
2. Smith, G. D., D. L. Crosby and P. A. Lewis (1996), 'Inguinal hernia and a single strenuous event', *Annals of The Royal College of Surgeons of England* 78 (4), pp. 367–8.

You can catch poison ivy from someone who has it

1. American Academy of Dermatology, 'Poison Ivy, Oak & Sumac', http://www.aad.org/public/publications/pamphlets/skin_poison. html (accessed 15 May 2008).

2. Lee, N. P., and E. R. Arriola (1999), 'Poison ivy, oak, and sumac dermatitis, *West Journal of Medicine* 171 (5–6), pp. 354–5.

Poinsettias are toxic

1. Krenzelok, E. P., T. D. Jacobsen and J. M. Aronis (1996), 'Poinsettia exposures have good outcomes … just as we thought', *American Journal of Emergency Medicine* 14 (7), pp. 671–4.
2. Paghat's Garden, 'Poinsettia pulcherrima', http://www.paghat.com/poinsettias.html (accessed 31 October 2008).
3. Stone, R. P., and W. J. Collins (1971), 'Euphorbia pulcherrima: toxicity to rats', *Toxicon* 9 (3), pp. 301–2.

You should put butter on a burn

1. American Academy of Pediatrics (2007), 'Parenting Corner Q&A: Burns', http://www.aap.org/publiced/BR_FireSafety_Burns.htm (accessed 21 May 2008).
2. Tiernan, E., and A. Harris (1993), 'Butter in the initial treatment of hot tar burn', *Burns* 19 (5), pp. 437–8.

If you have allergies, you should own only a short-haired or non-shedding dog

1. American Academy of Allergy, Asthma & Immunology, Allergy & Asthma Disease Management Center, 'Ask the Expert' (27 July 2006), http://www.asthmacasestudies.org/aadmc/ate/category.asp?cat=993 (accessed 15 May 2008).
2. Lindgren, S., et al. (1988), 'Breed-specific dog-dandruff allergens', *Journal of Allergy and Clinical Immunology* 82 (2), pp. 196–204.

3.   Ramadour, M., et al. (2005), 'Dog factor differences in Can f 1 allergen production', *Allergy* 60 (8), pp. 1060–64.

A dog's mouth is cleaner than a human's

1.   Callaham, M. (1988), 'Controversies in antibiotic choices for bite wounds', *Annals of Emergency Medicine* 17 (12), pp. 1321–30.
2.   Griego, R. D., et al. (1995), 'Dog, cat, and human bites: a review', *Journal of the American Academy of Dermatology* 33 (6), pp. 1019–29.
3.   Talan, D. A., et al. (1999), 'Bacteriologic analysis of infected dog and cat bites, for the Emergency Medicine Animal Bite Infection Study Group', *New England Journal of Medicine* 340 (2), pp. 85–92.

If you get stung by a bee sting, don't squeeze out the stinger

1.   American Academy of Pediatrics (1997), 'Speed overrides method in bee stinger removal', *AAP News* 13 (6), p. 16.
2.   Visscher, P. Kirk, Richard S. Vetter and Scott Camazine (1996), 'Removing bee stings', *The Lancet* 348 (9023), pp. 301–2.

If you get stung by a jellyfish, you should get someone to urinate on the sting

1.   Loten, C., et al. (2006), 'A randomised controlled trial of hot water (45° C) immersion versus ice packs for pain relief in bluebottle stings', *The Medical Journal of Australia* 184 (7), pp. 329–33.
2.   Little, Mark (2008), 'First aid for jellyfish stings: do we really know what we are doing?', *Emergency Medicine Australasia* 20 (1), pp. 78–80.

Citronella candles effectively repel mosquitoes

1. Fradin, Mark S., and John F. Day (2002), 'Comparative efficacy of insect repellents against mosquito bites', *New England Journal of Medicine* 347 (1), pp. 13–18.

2. Lindsay, L. R., et al. (1996), 'Evaluation of the efficacy of 3 per cent citronella candles and 5 per cent citronella incense for protection against field populations of Aedes mosquitoes', *Journal of the American Mosquito Control Association* 12 (2, Pt 1), pp. 293–4.

3. Roberts, J. R., and J. R. Reigart (2004), 'Does anything beat DEET?', *Pediatric Annals* 33 (7), pp. 443–53.

Mosquitoes that buzz by your ear don't bite

1. National Biological Information Infrastructure (US), 'Mosquito control and West Nile Virus: Mosquitoes 101', http://wildlifedisease.nbii.gov/westnilevirus/mosquitoes.html (accessed 21 November 2008).

2. Turpin, T. (2003), 'In mosquitoes, the lady is a vamp', http://www.agriculture.purdue.edu/agcomm/newscolumns/archives/OSL/2003/June/030612OSLhtml.htm (accessed 21 November 2008).

Green mucus indicates a sinus infection

1. Brook, I. (2003), 'Microbial dynamics of purulent nasopharyngitis in children', *International Journal of Pediatric Otorhinolaryngology* 67 (10), pp. 1047–53.

2. Mainous A. G., W. J. Hueston and C. Eberlein (1997), 'Colour of respiratory discharge and antibiotic use', *The Lancet* 350 (9084), p. 1077.

3. Todd, J. K., et al. (1984), 'Bacteriology and treatment of purulent

nasopharyngitis: a double-blind, placebo-controlled evaluation', *The Pediatric Infectious Disease Journal* 3 (3), pp. 226–32.

Using underarm antiperspirants causes breast cancer

1. Jemal, A., et al. (2006), 'Cancer statistics, 2006', *CA: A Cancer Journal for Clinicians* 56 (2), pp. 106–30.
2. Jones, J. (2000), 'Can rumors cause cancer?', *Journal of the National Cancer Institute* 92 (18), pp. 1469–71.
3. Mirick, D. K., S. Davis and D. B. Thomas (2002), 'Antiperspirant use and the risk of breast cancer', *Journal of the National Cancer Institute* 94 (20), pp. 1578–80.
4. Wisdom, J. P., M. Berlin and J. A. Lapidus (2005), 'Relating health policy to women's health outcomes', *Social Science & Medicine* 61 (8), pp. 1776–84.

Flu jabs can cause the flu

1. Blank, P. R., et al. (2008), 'Trends in influenza coverage rates in the United Kingdom over six seasons from 2001–2 to 2006–7', *Eurosurveillance* 13 (43), pp. 1–7.
2. Bridges, C. B., et al. (2000), 'Effectiveness and cost-benefit of influenza vaccination of healthy working adults: a randomized controlled trial', *Journal of the American Medical Association* 284 (13), pp. 1655–63.
3. Centers for Disease Control and Prevention (US), 'Key facts about seasonal flu vaccine', http://www.cdc.gov/flu/protect/keyfacts. htm (accessed 15 May 2008).
4. National Health Service, 'Having the vaccination', http://www. immunisation.nhs.uk/Vaccines/Flu/Having_the_vaccination (accessed 9 January 2009).
5. Nichol, K. L., et al. (1995), 'The effectiveness of vaccination against

influenza in healthy, working adults', *New England Journal of Medicine* 333 (14), pp. 889–93.

6.   Tosh, P. K., T. G. Boyce and G. A. Poland (2008), 'Flu myths: dispelling the myths associated with live attenuated influenza vaccine', *Mayo Clinic Proceedings* 83 (1), pp. 77–84.

You should stay awake if you've had a concussion

1.   American Academy of Family Physicians (1999), 'Concussion', http://www.aafp.org/afp/990901ap/990901e.html (accessed 21 May 2008).

2.   American Academy of Family Physicians (2006), 'Head injuries: what to watch for afterward', http://familydoctor.org/online/famdocen/home/common/brain/head/084.html (accessed 21 May 2008).

3.   Cassidy, J. D., et al. (2004), 'Incidence, risk factors and prevention of mild traumatic brain injury: results of the WHO Collaborating Centre Task Force on Mild Traumatic Brain Injury', *Journal of Rehabilitation Medicine* (Suppl. 43), pp. 28–60.

If you agree to donate your organs, doctors won't work as hard to save your life

1.   National Health Service (2008), 'Statistics', http://www.uktransplant.org.uk/ukt/statistics/statistics.jsp (accessed 9 January 2009).

2.   Schaeffner, E. S., et al. (2004), 'Knowledge and attitude regarding organ donation among medical students and doctors', *Transplantation* 77 (11), pp. 1714–18.

3.   United Network for Organ Sharing, 'Organ donation and transplantation', http://unos.org (accessed 15 May 2008).

If you use the highest SPF sun cream, you won't get burned

1. Diffey, Brian L. (2001), 'When should sunscreen be reapplied?', *Journal of the American Academy of Dermatology* 45 (6), pp. 882–5.
2. Foster R. D. (2002), '10 myths of skin care: someone's lying to you about how to avoid acne, wrinkles and skin cancer', http://findarticles.com/p/articles/mi_m1608/is_8_18/ai_89389712/pg_2 (accessed 11 June 2008).
3. Lademann, J., et al. (2004), 'Sunscreen application at the beach', *Journal of Cosmetic Dermatology* 3 (2), pp. 62–8.
4. Taylor, S. R. (2004), '"SunSmart Plus": the more informed use of sunscreens', *The Medical Journal of Australia* 180 (1), pp. 36–7.
5. Taylor, S., and B. Diffey (2002), 'Simple dosage guide for suncreams will help users', *British Medical Journal* 324 (7352), p.1526.

Vitamin C, echinacea and zinc will keep you from getting a cold

1. Douglas, R. M., et al. (2007), 'Vitamin C for preventing and treating the common cold', *Cochrane Database of Systematic Reviews* (3), CD000980.
2. Linde, K., et al. (2006), 'Echinacea for preventing and treating the common cold', *Cochrane Database of Systematic Reviews* (1), CD000530.
3. Schoop, R., et al. (2006), 'Echinacea in the prevention of induced rhinovirus colds: a meta-analysis', *Clinical Therapeutics* 28 (2), pp. 174–83.
4. Shah, S. A., et al. (2007), 'Evaluation of echinacea for the prevention and treatment of the common cold: a meta-analysis', *The Lancet Infectious Diseases* 7 (7), pp. 473–80.
5. Turner, R. B., et al. (2005), 'An evaluation of Echinacea angustifolia in experimental rhinovirus infections', *New England Journal of Medicine* 353 (4), pp. 341–8.
6. Alexander, T. H., and T. M. Davidson (2006), 'Intranasal zinc and

anosmia: the zinc-induced anosmia syndrome', *The Laryngoscope* 116, pp. 217–20.
7. Caruso, T. J., C. G. Prober and J. M. Gwaltney (2007), 'Treatment of naturally acquired common colds with zinc: a structured review', *Clinical Infectious Diseases* 45, pp. 569–74.

Breast milk can cure ear infections

1. Ip, S., et al. (2007), 'Breastfeeding and maternal and infant health outcomes in developed countries', *Evidence Report/Technology Assessment* (153), pp. 1–186.
2. Lubianca Neto, J. F., L. Hemb and D. B. Silva (2006), 'Systematic literature review of modifiable risk factors for recurrent acute otitis media in childhood', *Jornal de Pediatria* 82 (2), pp. 87–96.

Acupuncture doesn't really work

1. Buchbinder, R., et al. (2005), 'Shock wave therapy for lateral elbow pain', *Cochrane Database of Systematic Reviews* (4), CD003524.
2. Cheuk, D. K., et al. (2007), 'Acupuncture for insomnia', *Cochrane Database of Systematic Reviews* (3), CD005472.
3. Ezzo, J. M., et al. (2006), 'Acupuncture-point stimulation for chemotherapy-induced nausea or vomiting', *Cochrane Database of Systematic Reviews* (2), CD002285.
4. Furlan, A. D., et al. (2005), 'Acupuncture and dry-needling for low back pain', *Cochrane Database of Systematic Reviews* (1), CD001351.
5. Green, S., R. Buchbinder and S. Hetrick (2005), 'Acupuncture for shoulder pain', *Cochrane Database of Systematic Reviews* (2), CD005319.
6. Lee, A., and M. L. Done (2004), 'Stimulation of the wrist acupuncture point P6 for preventing postoperative nausea and vomiting', *Cochrane Database of Systematic Reviews* (3), CD003281.

7. Lim, B., et al. (2006), 'Acupuncture for treatment of irritable bowel syndrome', *Cochrane Database of Systematic Reviews* (4), CD005111.

8. Melchart, D., et al. (2001), 'Acupuncture for idiopathic headache', *Cochrane Database of Systematic Reviews* (1), CD001218.

9. Smith, C. A., and P. P. Hay (2005), 'Acupuncture for depression', *Cochrane Database of Systematic Reviews* (2), CD004046.

10. Smith, C. A., et al. (2006), 'Complementary and alternative therapies for pain management in labour', *Cochrane Database of Systematic Reviews* (4), CD003521.

11. Trinh, K.V., et al. (2006), 'Acupuncture for neck disorders', *Cochrane Database of Systematic Reviews* 3, CD004870.

12. Trinh, K., et al. (2007), 'Acupuncture for neck disorders', *Spine* 32 (2), pp. 236–43.

Men think about sex every seven seconds

1. Laumann, E. O. (1994), *The Social Organization of Sexuality: Sexual Practices in the United States* (Chicago, Ill.: University of Chicago Press).

2. The Kinsey Institute, 'Sexuality Information Links – FAQ', http://www.kinseyinstitute.org/resources/FAQ.html (accessed 19 May 2008).

Condoms protect you from all STDs

1. Steiner, M. J., and W. Cates, Jr (2006), 'Condoms and sexually transmitted infections', *New England Journal of Medicine* 354 (25), pp. 2642–3.

2. Wald, A., et al. (2005), 'The relationship between condom use and herpes simplex virus acquisition', *Annals of Internal Medicine* 143 (10), pp. 707–13.

3. Warner, L., et al. (2006), 'Condom use and risk of gonorrhea and chlamydia: a systematic review of design and measurement factors assessed in epidemiologic studies', *Sexually Transmitted Diseases* 33 (1), pp. 36–51.

Semen is loaded with calories

1. Kavanagh, J. P. (1985), 'Sodium, potassium, calcium, magnesium, zinc, citrate and chloride content of human prostatic and seminal fluid', *Journal of Reproduction and Fertility* 75 (1), pp. 35–41.
2. Keel, B. A. (2006), 'Within- and between-subject variation in semen parameters in infertile men and normal semen donors', *Fertility and Sterility* 85 (1), pp. 128–34.
3. Poland, M. L., et al. (1985), 'Variation of semen measures within normal men', *Fertility and Sterility* 44 (3), pp. 396–400.

Single people have much better sex lives than married people

1. Johnson, Anne M. (1994), *Sexual Attitudes and Lifestyles* (Oxford, England, and Boston, Mass.: Blackwell Scientific Publications).
2. Laumann, E. O. (1994), *The Social Organization of Sexuality: Sexual Practices in the United States* (Chicago, Ill.: University of Chicago Press).
3. National Opinion Research Center (US) (2006), 'American sexual behavior', http://www.norc.org/publications/american+sexual+behavior+2006.htm (accessed 20 May 2008).
4. Wellings, Kaye (1994), *Sexual Behaviour in Britain: The National Survey of Sexual Attitudes and Lifestyles* (London and New York: Penguin Books).

Women lose interest in sex after menopause

1. Hartmann, U., et al. (2004), 'Low sexual desire in midlife and older women: personality factors, psychosocial development, present sexuality', *Menopause* 11 (6, Pt 2), pp. 726–40.
2. Laumann, E. O. (1994), *The Social Organization of Sexuality: Sexual Practices in the United States* (Chicago, Ill.: University of Chicago Press).

Virgins don't have openings in their hymens

1. Acar, A., et al. (2007), 'The treatment of 65 women with imperforate hymen by a central incision and application of Foley catheter', *BJOG: An International Journal of Obstetrics & Gynaecology* 114 (11), pp. 1376–9.
2. Dane, C., et al. (2007), 'Imperforate hymen – a rare cause of abdominal pain: two cases and review of the literature', *Journal of Pediatric and Adolescent Gynecology* 20 (4), pp. 245–7.
3. Miller, R. J., and L. L. Breech (2008), 'Surgical correction of vaginal anomalies', *Clinical Obstetrics and Gynecology* 51 (2), pp. 223–36.

A doctor can tell whether you are a virgin or not

1. Adams, J. A., A. S. Botash and N. Kellogg (2004), Differences in hymenal morphology between adolescent girls with and without a history of consensual sexual intercourse', *Archives of Pediatrics and Adolescent Medicine* 158 (3), pp. 280–85.
2. Curtis, E., and C. San Lazaro (1999), 'Appearance of the hymen in adolescents is not well documented', *British Medical Journal* 318 (7183), p. 605.
3. Rogers, D. J., and M. Stark (1998), 'The hymen is not necessarily torn after sexual intercourse', *British Medical Journal* 317 (7155), p. 414.

You can't get pregnant using the 'pull out' method

1. Aytekin, N. T., et al. (2001), 'Family planning choices and some characteristics of coitus interruptus users in Gemlik, Turkey', *Women's Health Issues* 11 (5), pp. 442–7.
2. Bahadur, A., et al. (2008), 'Socio-demographic profile of women undergoing abortion in a tertiary centre', *Archives of Gynecology and Obstetrics* 278 (4), pp. 329–32.
3. Grady, W. R., M. D. Hayward and J. Yagi (1986), Contraceptive failure in the United States: estimates from the 1982 National Survey of Family Growth', *Family Planning Perspectives* 18 (5), pp. 200–209.
4. KidsHealth.org (2007), 'Withdrawal', http://www.kidshealth.org/PageManager.jsp?dn=familydoctor&article_set=10581&lic=44&cat_id=20018 (accessed 21 May 2008).

You can't get pregnant during your period

1. KidsHealth.org (2006), 'Can a girl get pregnant if she has sex during her period?', http://www.kidshealth.org/PageManager.jsp?dn=familydoctor&lic=44&cat_id=20015&article_set=20406&ps=209 (accessed 20 May 2008).

You can't get pregnant if you have sex in water

1. The National Campaign to Prevent Teen and Unplanned Pregnancy (US), 'Fact sheet', http://www.teenpregnancy.org/resources/reading/pdf/myths.pdf (accessed 11 June 2008).
2. Planned Parenthood of North Texas, Inc., 'Birth control', http://www.ppnt.org/sexual-health/birth-control/birth-control.html (accessed 11 June 2008).

Birth-control pills don't work as well if you're on antibiotics

1. Burroughs, K. E., and M. L. Chambliss (2000), 'Antibiotics and oral contraceptive failure', *Archives of Family Medicine* 9 (1), pp. 81–2.
2. Helms, S. E., et al. (1997), 'Oral contraceptive failure rates and oral antibiotics', *Journal of the American Academy of Dermatology* 36 (5, Pt 1), pp. 705–10.

You can't get pregnant when you're on the pill

1. KidsHealth.org (2007), 'Birth control pill', http://www.kidshealth. org/PageManager.jsp?dn=familydoctor&article_set=20395&lic= 44&cat_id=20018 (accessed 21 May 2008).

You're more likely to have a baby boy if you conceive in the middle of your cycle

1. France, J. T., et al. (1984), 'A prospective study of the preselection of the sex of offspring by timing intercourse relative to ovulation', *Fertility & Sterility* 41 (6), pp. 894–900.
2. Gray, R. H. (1991), 'Natural family planning and sex selection: fact or fiction?', *American Journal of Obstetrics & Gynecology* 165 (6, Pt 2), pp. 1982–4.
3. Mathews, F., P. J. Johnson and A. Neil (2008), 'You are what your mother eats: evidence for maternal preconception diet influencing foetal sex in humans', *Proceedings of the Royal Society of Biological Sciences* 275(1643), pp. 1661–8.
4. Simcock, B. W. (1985), 'Sons and daughters – a sex preselection study', *Medical Journal of Australia* 142 (10), pp. 541–2.

You can predict the sex of your baby without a doctor

1. Basso, O., and J. Olsen (2001), 'Sex ratio and twinning in women with hyperemesis or pre-eclampsia', *Epidemiology* 12 (6), pp. 747–9.
2. Costigan, K. A., H. L. Sipsma and J. A. DiPietro (2006), 'Pregnancy folklore revisited: the case of heartburn and hair', *Birth* 33 (4), pp. 311–14.
3. del Mar Melero-Montes, M., and H. Jick (2001), 'Hyperemesis gravidarum and the sex of the offspring', *Epidemiology* 12 (1), pp. 123–4.
4. Depue, R. H., et al. (1987), 'Hyperemesis gravidarum in relation to estradiol levels, pregnancy outcome, and other maternal factors: a seroepidemiologic study', *American Journal of Obstetrics & Gynecology* 156 (5), pp. 1137–41.
5. Druzin, M. L., J. M. Hutson and T. G. Edersheim (1986), 'Relationship of baseline fetal heart rate to gestational age and fetal sex', *American Journal of Obstetrics & Gynecology* 154 (5), pp. 1102–3.
6. Genuis, S., S. K. Genuis and W. C. Chang (1996), 'Antenatal fetal heart rate and "maternal intuition" as predictors of fetal sex', *Journal of Reproductive Medicine* 41 (6), pp. 447–9.
7. James, W. H. (1995), 'Sex ratios of offspring and the causes of placental pathology', *Human Reproduction* 10 (6), pp. 1403–6.
8. James, W. H. (2001), 'The associated offspring sex ratios and cause(s) of hyperemesis gravidarum', *Acta Obstetricia et Gynecologica Scandinavica* 80 (4), pp. 378–9.
9. James, W. H. (2004), 'The sex ratio of pregnancies complicated by hospitalisation for hyperemesis gravidarum', *BJOG: An International Journal of Obstetrics & Gynaecology* 111 (6), p. 636.
10. McKenna, D. S., et al. (2006), 'Gender-related differences in fetal heart rate during first trimester', *Fetal Diagnosis and Therapy* 21 (1), pp. 144–7.
11. Ostler, S., and A. Sun (1999), 'Fetal sex determination: the predictive value of 3 common myths', *Canadian Medical Association Journal* 161 (12), pp. 1525–6.
12. Perry, D. F., J. DiPietro and K. Costigan (1999), 'Are women carrying

"basketballs" really having boys? Testing pregnancy folklore', *Birth* 26 (3), pp. 172–7.

13. Petrie, B., and S. J. Segalowitz (1980), 'Use of fetal heart rate, other perinatal and maternal factors as predictors of sex', *Perceptual & Motor Skills* 50 (3, Pt 1), pp. 871–4.

14. Robles de Medina, P. G., et al. (2003), 'Fetal behaviour does not differ between boys and girls', *Early Human Development* 73 (1–2), pp. 17–26.

Twins skip a generation

1. Gilfillan C. P., et al. (1996), 'The control of ovulation in mothers of dizygotic twins', *Journal of Clinical Endocrinology & Metabolism* 81, pp. 1557–62.

2. Hoekstra C. Z., et al. (2008), 'Dizygotic twinning', *Human Reproduction Update* 14, pp. 37–47.

3. Lambalk C. B. (2001), 'Is there a role for follicle-stimulating-hormone receptor in familial dizygotic twinning?', *The Lancet* 357, pp. 735–6.

4. Lambalk C. B., and J. Schoemaker (1997), 'Hypothetical risks of twinning in the natural menstrual cycle', *European Journal of Obstetrics & Gynecology and Reproductive Biology* 75, pp.1–4.

5. O'Connor, A. (2 October 2007), 'The claim: twins always skip a generation', *New York Times*, http://www.nytimes.com/2007/10/02/health/02real.html (accessed 13 March 2009).

Flying on a plane is dangerous for your unborn baby

1. American College of Obstetricians and Gynecologists, Committee on Obstetric Practice (2001), 'Committee opinion: number 264. Air travel during pregnancy', *Obstetrics & Gynecology* 98 (6), pp. 1187–8.

2.   Daniell, W. E., T. L. Vaughan and B. A. Millies (1990), 'Pregnancy outcomes among female flight attendants', *Aviation, Space, and Environmental Medicine* 61 (9), pp. 840–44.

3.   Freeman, M., et al. (2004), 'Does air travel affect pregnancy outcome?', *Archives of Gynecology & Obstetrics* 269 (4), pp. 274–7.

4.   Friedberg, W., et al. (1989), 'Galactic cosmic radiation exposure and associated health risks for air carrier crewmembers', *Aviation, Space, and Environmental Medicine* 60 (11), pp. 1104–8.

5.   Morrell, S., R. Taylor and D. Lyle (1997), 'A review of health effects of aircraft noise', *Australian and New Zealand Journal of Public Health* 21 (2), pp. 221–36.

Bed rest prevents pre-term labour

1.   Sosa, C., et al. (2004), 'Bed rest in singleton pregnancies for preventing preterm birth', *Cochrane Database of Systematic Reviews* (1), CD003581.

Baby Einstein® makes your baby smarter

1.   Nyhan P. (8 August 2007), 'Videos won't make baby smart', *Seattle Post-Intelligence*.

2.   Thakkar, R. R., M. M. Garrison, and D. A. Christakis (2006), 'A systematic review for the effects of television viewing by infants and preschoolers', *Pediatrics* 118 (5), pp. 2025–31.

3.   Toppo G. (3 April 2007), 'Report puts a pacifier on "smarter baby" debate', *USA TODAY*.

4.   Zimmerman, F. J., and D. A. Christakis (2005), 'Children's television viewing and cognitive outcomes: a longitudinal analysis of national data', *Archives of Pediatrics and Adolescent Medicine* 159 (7), pp. 619–25.

5.   Zimmerman, F. J., and D. A. Christakis (2007), 'Associations

between content types of early media exposure and subsequent attentional problems', *Pediatrics* 120 (5), pp. 986–92.

6. Zimmerman, F.J., D.A. Christakis and A.N. Meltzoff (2007), 'Television and DVD/video viewing in children younger than 2 years', *Archives of Pediatrics and Adolescent Medicine* 161 (5), pp. 473–9.

Adding cereal to your baby's diet helps him sleep longer

1. Crocetti, M., R. Dudas and S. Krugman (2004), 'Parental beliefs and practices regarding early introduction of solid foods to their children', *Clinical Pediatrics (Philadelphia)* 43 (6), pp. 541–7.

2. Macknin, M. L., S.V. Medendorp and M. C. Maier (1989), 'Infant sleep and bedtime cereal', *American Journal of Diseases of Children* 143 (9), pp. 1066–8.

3. Nikolopoulou, M., and I. St James-Roberts (2003), 'Preventing sleeping problems in infants who are at risk of developing them', *Archives of Disease in Childhood* 88 (2), pp. 108–11.

4. Pinilla, T., and L. L. Birch (1993), 'Help me make it through the night: behavioral entrainment of breast-fed infants' sleep patterns', *Pediatrics* 91 (2), pp. 436–44.

5. Robertson, R. M. (1974), 'Solids and "sleeping through"', *British Medical Journal* 1 (5900), p. 200.

6. St James-Roberts, I., et al. (2001), 'Use of a behavioural programme in the first 3 months to prevent infant crying and sleeping problems', *Journal of Paediatrics and Child Health* 37, pp. 289–97.

Propping a baby upright helps with reflux

1. Carroll, A. E., M. M. Garrison and D. A. Christakis (2002), 'A systematic review of non-pharmacological and non-surgical therapies for gastroesophageal reflux in infants', *Archives of Pediatrics and Adolescent Medicine* 156 (2), pp. 109–13.

2. Chao, H. C., and Y. Vandenplas (2007), 'Effect of cereal-thickened formula and upright positioning on regurgitation, gastric empty-ing, and weight gain in infants with regurgitation', *Nutrition* 23 (1), pp. 23–8.

3. Craig, W. R., et al. (2004), 'Metoclopramide, thickened feedings, and positioning for gastro-oesophageal reflux in children under two years', *Cochrane Database of Systematic Reviews* (4), CD003502.

Teething can cause a fever

1. Barlow, B. S., M. J. Kanellis and R. L. Slayton (2002), 'Tooth erup-tion symptoms: a survey of parents and health professionals', *Jour-nal of Dentistry for Children* 69 (2), pp. 148–50.

2. Frank, J., and J. Drezner (2001), 'Is teething in infants associated with fever or other symptoms?', *The Journal of Family Practice* 50 (3), p. 257.

3. Sarrell, E. M., et al. (2005), 'Parents' and medical personnel's beliefs about infant teething', *Patient Education and Counseling* 57 (1), pp. 122–5.

4. Wake, M., K. Hesketh and J. Lucas (2000), 'Teething and tooth eruption in infants: a cohort study', *Pediatrics* 106 (6), pp. 1374–9.

It's possible to spoil a baby

1. American Academy of Pediatrics (2007), 'Parenting Corner Q&A: Separation Anxiety', http://www.aap.org/publiced/BK0_SeparationAnxiety.htm (accessed 20 May 2008).

2. Baby Center, 'Should I worry about spoiling my baby?', http://www.babycenter.com/404_should-i-worry-about-spoiling-my-baby_3446.bc (accessed 20 May 2008).

3. Buchholz, E. S. (1997), *The Call of Solitude: Alonetime in a World of Attachment* (New York: Simon and Schuster).

4. DrGreene.com, 'Spoiling a Baby', http://www.drgreene.com/21_5.html (accessed 20 May 2008).

5. Pascoe, J. M., and R. Solomon (1994), 'Prenatal correlates of indigent mothers' attitudes about spoiling their young infants: a longitudinal study', *Journal of Developmental and Behavioral Pediatrics* 15 (5), pp. 367–9.

6. University of Minnesota Extension, 'Comforting your baby doesn't mean spoiling', http://www.extension.umn.edu/info-u/babies/BE328.html (accessed 20 May 2008).

7. Wilson, A. L., D. B. Witzke and A. Volin (1981), 'What it means to "spoil" a baby: parents' perception', *Clinical Pediatrics (Philadelphia)* 20 (12), pp. 798–802.

Women who are breastfeeding can drink alcohol

1. Breslow, R. A., et al. (2007), 'Alcohol consumption among breastfeeding women', *Breastfeeding Medicine* 2 (3), pp. 152–7.

2. Liston, J. (1998), 'Breastfeeding and the use of recreational drugs – alcohol, caffeine, nicotine and marijuana', *Breastfeeding Review* 6 (2), pp. 27–30.

3. Little, R. E., et al. (1989), 'Maternal alcohol use during breastfeeding and infant mental and motor development at one year', *New England Journal of Medicine* 321 (7), pp. 425–30.

4. Little, R. E., K. Northstone and J. Golding (2002), 'Alcohol, breastfeeding, and development at 18 months', *Pediatrics* 109 (5), e72.

5. Mennella, J. (2001), 'Alcohol's effect on lactation', *Alcohol Research & Health* 25 (3), pp. 230–34.

6. Mennella, J. A., and G. K. Beauchamp (1991), 'The transfer of alcohol to human milk: effects on flavor and the infant's behavior', *New England Journal of Medicine* 325 (14), pp. 981–5.

7. Mennella, J. A., and C. J. Gerrish (1998), 'Effects of exposure to alcohol in mother's milk on infant sleep', *Pediatrics* 101 (5), e2.

Over-the-counter cold and cough medicines are safe for babies and toddlers

1. Bhatt-Mehta, V. (2004), 'Over-the-counter cough and cold medicines: should parents be using them for their children?', *The Annals of Pharmacotherapy* 38 (11), pp. 1964–6.

2. Chang, A. B., et al. (2006), 'Cough in children: definitions and clinical evaluation', *The Medical Journal of Australia* 184 (8), pp. 398–403.

3. Chang, A. B., J. Peake and M. S. McElrea (2008), 'Anti-histamines for prolonged non-specific cough in children', *Cochrane Database of Systematic Reviews* (2), CD005604.

4. Committee on Drugs (1997), 'Use of codeine- and dextromethorphan-containing cough remedies in children', *Pediatrics* 99 (6), pp. 918–20.

5. 'Infant deaths associated with cough and cold medications – two states, 2005' (2007), *Morbidity and Mortality Weekly Report* 56 (1), pp. 1–4.

6. Food and Drug Administration (US), Center for Drug Evaluation and Research (2008), 'Public health advisory: nonprescription cough and cold medicine use in children', http://www.fda.gov/cder/drug/advisory/cough_cold_2008.htm (accessed 11 June 2008).

7. 'Sale of infant cough medicines splutters to a halt' (editorial) (2008), *The Lancet* 371 (9619), p. 1138.

8. Sharfstein, J. M., M. North and J. R. Serwint (2007), Over the counter but no longer under the radar – pediatric cough and cold medications', *New England Journal of Medicine* 357 (23), pp. 2321–4.

9. Smith, S. M., K. Schroeder and T. Fahey (2008), 'Over-the-counter medications for acute cough in children and adults in ambulatory settings', *Cochrane Database of Systematic Reviews* (1), CD001831.

Walkers help your baby learn to walk earlier

1. Burrows, P., and P. Griffiths (2002), 'Do baby walkers delay onset of walking in young children?', *British Journal of Community Nursing* 7 (11), pp. 581–6.
2. Crouchman, M. (1986), 'The effects of baby walkers on early locomotor development', *Developmental Medicine & Child Neurology* 28 (6), pp. 757–61.
3. Kendrick, D., et al. (2003), 'Baby walkers – health visitors' current practice, attitudes and knowledge', *Journal of Advanced Nursing* 43 (5), pp. 488–95.
4. Khambalia, A., et al. (2006), 'Risk factors for unintentional injuries due to falls in children aged 0–6 years: a systematic review', *Injury Prevention* 12 (6), pp. 378–81.
5. LeBlanc, J. C., et al. (2006), 'Home safety measures and the risk of unintentional injury among young children: a multicentre case-control study', *Canadian Medical Association Journal* 175 (8), pp. 883–7.
6. Pin, T., B. Eldridge and M. P. Galea (2007), 'A review of the effects of sleep position, play position, and equipment use on motor development in infants', *Developmental Medicine & Child Neurology* 49 (11), pp. 858–67.
7. Rhodes, K., D. Kendrick and J. Collier (2003), Baby walkers: paediatricians' knowledge, attitudes, and health promotion', *Archives of Disease in Childhood* 88 (12), pp. 1084–5.
8. Ridenour, M. V. (1982), 'Infant walkers: developmental tool or inherent danger', *Perceptual & Motor Skills* 55 (3, Pt 2), pp. 1201–2.
9. Siegel, A. C., and R. V. Burton (1999), 'Effects of baby walkers on motor and mental development in human infants', *Journal of Developmental and Behavioral Pediatrics* 20 (5), pp. 355–61.

The iron in baby formula causes constipation

1. Harrod-Wild, K. (2007), 'Does childhood nutrition matter?', *Journal of Family Health Care* 17 (3), pp. 89–91.
2. Heresi, G., et al. (1995), 'Effect of supplementation with an iron-fortified milk on incidence of diarrhea and respiratory infection in urban-resident infants', *Scandinavian Journal of Infectious Diseases* 27 (4), pp. 385–9.
3. Malacaman, E. E., et al. (1985), 'Effect of protein source and iron content of infant formula on stool characteristics', *Journal of Pediatric Gastroenterology and Nutrition* 4 (5), pp. 771–3.
4. Pizarro, F., et al. (1991), 'Iron status with different infant feeding regimens: relevance to screening and prevention of iron deficiency', *Journal of Pediatrics* 118 (5), pp. 687–92.
5. Scariati, P. D., et al. (1997), 'Risk of diarrhea related to iron content of infant formula: lack of evidence to support the use of low-iron formula as a supplement for breastfed infants', *Pediatrics* 99 (3), e2.
6. Singhal, A., et al. (2000), 'Clinical safety of iron-fortified formulas', *Pediatrics* 105 (3), e38.

Babies need water when it is hot outside

1. Bruce, R. C., and R. M. Kliegman (1997), 'Hyponatremic seizures secondary to oral water intoxication in infancy: association with commercial bottled drinking water', *Pediatrics* 100 (6), e4.
2. 'Hyponatremic seizures among infants fed with commercial bottled drinking water – Wisconsin, 1993' (1994), *Morbidity and Mortality Weekly Report* 43 (35), pp. 641–3.
3. Levallois, P., et al. (2008), 'Drinking water intake by infants living in rural Quebec (Canada)', *Science of the Total Environment* 397 (1–3), pp. 82–5.
4. Manz, F. (2007), 'Hydration in children', *Journal of the American College of Nutrition* 26 (5, suppl.), pp. 562–9S.

Sugar makes kids hyperactive

1. Hoover, D. W., and R. Milich (1994), 'Effects of sugar ingestion expectancies on mother–child interactions', *Journal of Abnormal Child Psychology* 22 (4), pp. 501–15.
2. Kinsbourne, M. (1994), 'Sugar and the hyperactive child', *New England Journal of Medicine* 330 (5), pp. 355–6.
3. Krummel, D. A., F. H. Seligson and H. A. Guthrie (1996), 'Hyperactivity: is candy causal?', *Critical Reviews in Food Science and Nutrition* 36 (1–2), pp. 31–47.
4. Wolraich, M. L., et al. (1994), 'Effects of diets high in sucrose or aspartame on the behavior and cognitive performance of children', *New England Journal of Medicine* 330 (5), pp. 301–7.

Eat your spinach to grow strong like Popeye

1. Cardwell, G. (2005), 'Spinach is a good source of what?', *The Skeptic* 25 (2), pp. 31–3.
2. Hamblin, T. J. (1981), 'Fake', *British Medical Journal* 283 (6307), pp. 1671–4.
3. Rogers, J. (1990), *What Food Is That?: And How Healthy Is It?* (Sydney: Lansdowne Publishing Pty Ltd).
4. Rutzke, C. J., et al. (2004), 'Bioavailability of iron from spinach using an in vitro/human Caco-2 cell bioassay model', *Habitation (Elmsford)* 10 (1), pp. 7–14.
5. Singh, G., A. Kawatra and S. Sehgal (2001), 'Nutritional composition of selected green leafy vegetables, herbs and carrots', *Plant Foods for Human Nutrition* 56 (4), pp. 359–64.
6. Zhang, D., D. G. Hendricks and A. W. Mahoney (1989), 'Bioavailability of total iron from meat, spinach (Spinacea oleracea L.) and meat-spinach mixtures by anaemic and non-anaemic rats', *British Journal of Nutrition* 61 (2), pp. 331–43.

Chewing gum stays in your stomach for seven years

1. Edgar, W. M. (1998), 'Sugar substitutes, chewing gum and dental caries – a review', *British Dental Journal* 184 (1), pp. 29–32.
2. Imfeld, T. (1999), 'Chewing gum – facts and fiction: a review of gum-chewing and oral health', *Critical Reviews in Oral Biology & Medicine* 10 (3), pp. 405–19.
3. Milov, D. E., et al. (1998), 'Chewing gum bezoars of the gastrointestinal tract', *Pediatrics* 102 (2), e22.

Eating turkey makes you sleepy

1. Barrett, P. R., J. A. Horne and L. A. Reyner (2005), 'Early evening low alcohol intake also worsens sleepiness-related driving impairment', *Human Psychopharmacology* 20 (4), pp. 287–90.
2. Canadian Food Inspection Agency (1996), *Guide to Food Labeling and Advertising*, http://www.inspection.gc.ca/english/fssa/labeti/guide/toce.shtml (accessed 6 March 2009).
3. Food and Drug Administration (US), Center for Food Safety and Applied Nutrition (2001), 'Information Paper on L-tryptophan and 5-hydroxy-L-tryptophan', http://vm.cfsan.fda.gov/~dms/ds-tryp1.html (accessed 6 March 2009).
4. Holt, S. H., et al. (1999), 'The effects of high-carbohydrate vs high-fat breakfasts on feelings of fullness and alertness, and subsequent food intake', *International Journal of Food Sciences and Nutrition* 50 (1), pp. 13–28.
5. Hoost, U., et al. (1996), 'Haemodynamic effects of eating: the role of meal composition', *Clinical Science* 90 (4), pp. 269–76.
6. Hudson, C., et al. (2005), 'Protein source tryptophan versus pharmaceutical grade tryptophan as an efficacious treatment for chronic insomnia', *Nutritional Neuroscience* 8 (2), pp. 121–7.
7. Lenard, N. R., and A. J. Dunn (2005), 'Mechanisms and significance

of the increased brain uptake of tryptophan', *Neurochemical Research* 30 (12), pp. 1543–8.

8. Lloyd, H. M., M. W. Green and P. J. Rogers (1994), 'Mood and cognitive performance effects of isocaloric lunches differing in fat and carbohydrate content', *Physiology & Behavior* 56 (1), pp. 51–7.

9. Paredes, S. D., et al. (2007), 'Tryptophan increases nocturnal rest and affects melatonin and serotonin serum levels in old ringdove', *Physiology & Behavior* 90 (4), pp. 576–82.

10. Paz, A., and E. M. Berry (1997), 'Effect of meal composition on alertness and performance of hospital night-shift workers. Do mood and performance have different determinants?', *Annals of Nutrition and Metabolism* 41 (5), pp. 291–8.

11. Platten, M., et al. (2005), 'Treatment of autoimmune neuroinflammation with a synthetic tryptophan metabolite', *Science* 310 (5749), pp. 850–55.

12. Van Reen, E., O. G. Jenni and M. A. Carskadon (2006), 'Effects of alcohol on sleep and the sleep electroencephalogram in healthy young women', *Alcoholism: Clinical and Experimental Research* 30 (6), pp. 974–81.

13. Wells, A. S., and N. W. Read (1996), 'Influences of fat, energy, and time of day on mood and performance', *Physiology & Behavior* 59 (6), pp. 1069–76.

14. Wells, A. S., et al. (1998), 'Effects of meals on objective and subjective measures of daytime sleepiness', *Journal of Applied Physiology* 84 (2), pp. 507–15.

15. Wurtman, R. J., et al. (2003), 'Effects of normal meals rich in carbohydrates or proteins on plasma tryptophan and tyrosine ratios', *American Journal of Clinical Nutrition* 77 (1), pp. 128–32.

Milk makes you phlegmy

1. Arney, W. K., and C. B. Pinnock (1993), 'The milk mucus belief: sensations associated with the belief and characteristics of believers', *Appetite* 20 (1), pp. 53–60.
2. Lee, C., and A. J. Dozor (2004), 'Do you believe milk makes mucus?', *Archives of Pediatrics and Adolescent Medicine* 158 (6), pp. 601–3.
3. Pinnock, C. B., et al. (1990), 'Relationship between milk intake and mucus production in adult volunteers challenged with rhinovirus-2', *American Review of Respiratory Disease* 141 (2), pp. 352–6.
4. Pinnock, C. B., and W. K. Arney (1993), 'The milk-mucus belief: sensory analysis comparing cow's milk and a soy placebo', *Appetite* 20 (1), pp. 61–70.
5. Wuthrich, B., et al. (2005), 'Milk consumption does not lead to mucus production or occurrence of asthma', *Journal of the American College of Nutrition* 24 (6, suppl.), pp. 547–55S.

Eating bananas attracts mosquitoes, while eating garlic repels them

1. Fradin, Mark S. (1998), 'Mosquitoes and mosquito repellents: a clinician's guide', *Annals of Internal Medicine* 128 (11), pp. 931–40.

At a picnic, you should avoid foods with mayonnaise

1. Food and Drug Administration (US), Center for Food Safety and Applied Nutrition (2007), 'Eating outdoors: handling food safely', http://www.cfsan.fda.gov/~dms/fssummer.html (accessed 21 November 2008).
2. Smittle, R. B. (2000), 'Microbiological safety of mayonnaise, salad dressings, and sauces produced in the United States: a review', *Journal of Food Protection* 63 (8), pp. 1144–53.

Eating grapefruit burns calories

1. Ballard, Tasha L. P., et al. (2006), 'Naringin does not alter caffeine pharmacokinetics, energy expenditure, or cardiovascular haemo-dynamics in humans following caffeine consumption', *Clinical and Experimental Pharmacology and Physiology* 33 (4), pp. 310–14.

2. Cunningham, Eleese, and Wendy Marcason (2001), 'Is it possible to burn calories by eating grapefruit or vinegar?', *Journal of the American Dietetic Association* 101 (10), p. 1198.

3. de Jonge, L., and G. A. Bray (1997), 'The thermic effect of food and obesity: a critical review', *Obesity* 5 (6), pp. 622–31.

4. Duyff, Roberta Larson (1999), *Food Folklore: Tales and Truths About What We Eat* (Minneapolis, Minn.: Chronimed).

5. Fujioka, K., et al. (2006), 'The effects of grapefruit on weight and insulin resistance: relationship to the metabolic syndrome', *Journal of Medicinal Food* 9 (1), pp. 49–54.

6. Sanders, T. A., R. Woolfe and E. Rantzen (1990), 'Controlled evaluation of slimming diets: use of television for recruitment', *Lancet* 336 (8720), pp. 918–20.

7. Stump, A. L., T. Mayo and A. Blum (2006), 'Management of grapefruit–drug interactions', *American Family Physician* 74 (4), pp. 605–8.

Eating at night makes you fat

1. Andersson, I., and S. Rossner (1996), 'Meal patterns in obese and normal weight men: the "Gustaf" study', *European Journal of Clinical Nutrition* 50 (10), pp. 639–46.

2. Berteus Forslund, H., et al. (2002), 'Meal patterns and obesity in Swedish women – a simple instrument describing usual meal types, frequency and temporal distribution', *European Journal of Clinical Nutrition* 56 (8), pp. 740–47.

3. Consoli, A., et al. (1981), '[Effect of scheduling of meal times on

the circadian rhythm of energy expenditure]', *Bollettino della Società italiana di biologia sperimentale* 57 (23), pp. 2322–4.

4. Dubois, L., et al. (2009), 'Breakfast skipping is associated with differences in meal patterns, macronutrient intakes and overweight among pre-school children', *Public Health Nutrition* 12, pp. 19–28.

5. Howarth, N. C., et al. (2007), 'Eating patterns and dietary composition in relation to BMI in younger and older adults', *International Journal of Obesity* 31 (4), pp. 675–84.

6. Sjoberg, A., et al. (2003), 'Meal pattern, food choice, nutrient intake and lifestyle factors in The Goteborg Adolescence Study', *European Journal of Clinical Nutrition* 57 (12), pp. 1569–78.

You should drink at least eight glasses of water a day

1. Anonymous (2001), 'Water, water everywhere', *HealthNews* 7 (2), p. 3.

2. Associated Content (2007), 'Woman dead after radio station's water-drinking contest', http://www.associatedcontent.com/article/124583/woman_dead_after_radio_stations_waterdrinking.html (accessed 21 May 2008).

3. Clarke, J. (16 July 2005), 'At your table: water', *The Times*.

4. Department of Agriculture (US), National Technical Information Service (1994–1996, 1998; 2000), 'Continuing Survey of Food Intakes by Individuals (CSFII) 1994–1996' and 'Diet and Health Knowledge Survey, 2000'.

5. Grandjean, A. C., et al. (2000), 'The effect of caffeinated, non-caffeinated, caloric and non-caloric beverages on hydration', *Journal of the American College of Nutrition* 19 (5), pp. 591–600.

6. National Academy of Sciences (US), National Research Council, Food and Nutrition Board (1945), 'Recommended Dietary Allowances'.

7. Snopes.com, 'Eight glasses', http://www.snopes.com/medical/myths/8glasses.asp (accessed 11 June 2008).

8. Stare, F. J., and M. McWilliams (1974), *Nutrition for Good Health* (Fullerton, Calif.: Plycon).

9. Valtin, H. (2002), '"Drink at least eight glasses of water a day." Really? Is there scientific evidence for "8 x 8"?', *American Journal of Physiology – Regulatory, Integrative and Comparative Physiology* 283 (5), pp. 993–1004.

If you are thirsty, you are already dehydrated

1. Kratz, A., and K. B. Lewandrowski (1998), 'Case records of the Massachusetts General Hospital. Weekly clinicopathological exercises. Normal reference laboratory values', *New England Journal of Medicine* 339 (15), pp. 1063–72.

2. Negoianu, D., and S. Goldfarb (2008), 'Just add water', *Journal of the American Society of Nephrology* 19(6), pp. 1041–3.

3. Phillips, P. A., et al. (1984), 'Body fluid changes, thirst and drinking in man during free access to water', *Physiology & Behavior* 33, pp. 357–63.

4. Robertson, G. L. (1991), 'Disorders of Thirst in Man', in D. L. Ramsay and D. Booth (eds.), *Thirst: Physiological and Psychological Aspects* (London: Springer-Verlag).

5. Thompson, C. J., et al. (1986), 'The osmotic thresholds for thirst and vasopressin release are similar in healthy man', *Clinical Science* 71 (6), pp. 651–6.

6. Valtin, H. (2002), '"Drink at least eight glasses of water a day." Really? Is there scientific evidence for "8 x 8"?', *American Journal of Physiology – Regulatory, Integrative and Comparative Physiology* 283 (5), pp. 993–1004.

7. Valtin, H., and J. A. Schafer (1995), *Renal Function Mechanisms Preserving Fluid and Solute Balance in Health*, 3rd edn (Boston, Mass.: Little, Brown & Co.).

8. Weinberg, A. D., and K. L. Minaker (1995), 'Dehydration. Evaluation and management in older adults. Council on Scientific Affairs,

American Medical Association', *Journal of the American Medical Association* 274 (19), pp. 1552–6.

Caffeinated beverages are dehydrating

1. Armstrong, L. E. (2002), 'Caffeine, body fluid-electrolyte balance, and exercise performance', *International Journal of Sport Nutrition and Exercise Metabolism* 12 (2), pp. 189–206.
2. Armstrong, L. E., et al. (2005), 'Fluid, electrolyte, and renal indices of hydration during 11 days of controlled caffeine consumption', *International Journal of Sport Nutrition and Exercise Metabolism* 15 (3), pp. 252–65.
3. Armstrong, L. E., et al. (2007), 'Caffeine, fluid-electrolyte balance, temperature regulation, and exercise-heat tolerance', *Exercise and Sport Sciences Reviews* 35 (3), pp. 135–40.
4. Eddy, Nathan B., and Ardrey W. Downs (1928), 'Tolerance and Cross-Tolerance in the Human Subject to the Diuretic Effect of Caffeine, Theobromine and Theophylline', *The Journal of Pharmacology and Experimental Therapeutics* 33 (2), pp. 167–74.
5. Fiala, K. A., D. J. Casa and M. W. Roti (2004), 'Rehydration with a caffeinated beverage during the nonexercise periods of 3 consecutive days of 2-a-day practices', *International Journal of Sport Nutrition and Exercise Metabolism* 14 (4), pp. 419–29.
6. Grandjean, A. C., et al. (2000), 'The effect of caffeinated, non-caffeinated, caloric and non-caloric beverages on hydration', *Journal of the American College of Nutrition* 19 (5), pp. 591–600.
7. Maughan, R. J., and J. Griffin (2003), 'Caffeine ingestion and fluid balance: a review', *Journal of Human Nutrition and Dietetics* 16 (6), pp. 411–20.
8. Neuhauser, B., et al. (1997), 'Coffee consumption and total body water homeostasis as measured by fluid balance and bioelectrical impedance analysis', *Annals of Nutrition and Metabolism* 41 (1), pp. 29–36.

You can cure a hangover with ...

1. Jung, T. W., et al. (2006), 'Rosiglitazone relieves acute ethanol-induced hangover in Sprague-Dawley rats', *Alcohol* 41(3), pp. 231–5.
2. McGregor, N. R. (2007), 'Pueraria lobata (Kudzu root) hangover remedies and acetaldehyde-associated neoplasm risk', *Alcohol* 41(7), pp. 469–78.
3. Pittler, M. H., J. C. Verster and E. Ernst (2005), 'Interventions for preventing or treating alcohol hangover: systematic review of randomised controlled trials', *British Medical Journal* 331(7531), pp. 1515–18.
4. Venkataranganna, M. V., et al. (2008), 'Pharmacodynamics & toxicological profile of PartySmart, a herbal preparation for alcohol hangover in Wistar rats', *Indian Journal of Medical Research* 127 (5), pp. 460–66.

Beer before liquor, never been sicker

1. Chacko, D. (17 April 2006), 'Beer before liquor?', *Daily Trojan*.
2. Roberts, C., and S. P. Robinson (2007), 'Alcohol concentration and carbonation of drinks: the effect on blood alcohol levels', *Journal of Forensic and Legal Medicine* 14 (7), pp. 398–405.
3. Van Dusen, A. (12 March 2007), 'Alcohol and hangover myths exposed', Forbes.com.

If you pick up food within five seconds of it hitting the floor, it's safe to eat

1. Dawson, P., et al. (2007), 'Residence time and food contact time effects on transfer of *Salmonella* Typhimurium from tile, wood and carpet: testing the five-second rule', *Journal of Applied Microbiology* 102 (4), pp. 945–53.

2. De Cesare, A., et al. (2003), 'Survival and persistence of *Campylobacter* and *Salmonella* species under various organic loads on food contact surfaces', *Journal of Food Protection* 66 (9), pp. 1587–94.

3. Moore, G., I. S. Blair, and D. A. McDowell (2007), Recovery and transfer of *Salmonella* Typhimurium from four different domestic food contact surfaces', *Journal of Food Protection* 70 (10), pp. 2273–80.

4. Rohrer, C. A., et al. (2003), 'Transfer efficiencies of pesticides from household flooring surfaces to foods', *Journal of Exposure Analysis and Environmental Epidemiology* 13 (6), pp. 454–64.

You can chew gum instead of brushing your teeth

1. Addy, M., E. Perriam and A. Sterry (1982), 'Effects of sugared and sugar-free chewing gum on the accumulation of plaque and debris on the teeth', *Journal of Clinical Periodontology* 9 (4), pp. 346–54.

2. Ainamo, J., et al. (1979), 'Plaque growth while chewing sorbitol and xylitol simultaneously with sucrose flavored gum', *Journal of Clinical Periodontology* 6 (6), pp. 397–406.

3. Anderson, G. B., et al. (1990), 'Effects of zirconium silicate chewing gum on plaque and gingivitis', *Quintessence International* 21 (6), pp. 479–89.

4. Edgar, W. M. (1998), 'Sugar substitutes, chewing gum and dental caries – a review', *British Dental Journal* 184 (1), pp. 29–32.

5. Fletcher, J. (2004), *The Search for Nefertiti: The True Story of a Remarkable Discovery* (New York: W. Morrow).

6. Imfeld, T. (1999), 'Chewing gum – facts and fiction: a review of gum-chewing and oral health', *Critical Reviews in Oral Biology & Medicine* 10 (3), pp. 405–19.

7. Kleber, C. J., and M. S. Putt (1986), 'Plaque removal by a chewing gum containing zirconium silicate', *Compendium of Continuing Education in Dentistry* 7 (9), pp. 681–5.

8. Simpson, William Kelly, and Robert Kriech Ritner (2003), *The*

Literature of Ancient Egypt: An Anthology of Stories, Instructions, and Poetry, 3rd edn (New Haven, Conn., and London: Yale University Press).

You should wait an hour after eating before you go swimming

1. Engel, Peter, and Merrit Malloy (1993), *Old Wives' Tales: The Lowdown on Everyday Myths* (New York: St Martin's Press).
2. Snopes.com, 'Wait an hour after eating to swim', http://www.snopes.com/oldwives/hourwait.asp (accessed 21 May 2008).
3. Stoppler, M. (2008), 'Debunking summer health myths', http://www.medicinenet.com/script/main/art.asp?articlekey=47368 (accessed 21 May 2008).

It's safe to double-dip

1. McGee, H. (30 January 2008), 'Dip once or dip twice?', *New York Times*, http://www.nytimes.com/2008/01/30/dining/30curious.html?_r=3&pagewanted=print&oref=slogin&oref=slogin&oref=slogin (accessed 13 May 2008).
2. Trevino J., et al., Dept of Food Science & Human Nutrition, Clemson University, 'Double-dipping does transfer bacteria: George was wrong', http://www.clemson.edu/foodscience/PDF%20Downloads/Double-Dipping%20Does%20Transfer%20Bacteria.pdf (accessed May 13, 2008).

The fluoride in your water is dangerous

1. American Dental Association (2005), 'Fluoride & fluoridation: fluoridation facts introduction', http://www.ada.org/public/topics/fluoride/facts/index.asp (accessed 16 May 2008).

2. BBC News (6 October 2000), 'Fluoride in water "benefits health"', http://news.bbc.co.uk/2/hi/health/957372.stm (accessed 9 January 2009).

3. Centers for Disease Control and Prevention (US) (2000), 'Achievements in public health, 1900–1999: fluoridation of drinking water to prevent dental caries', *Journal of the American Medical Association* 283 (10), pp. 1283–6.

4. Horowitz, H. S. (1996), 'The effectiveness of community water fluoridation in the United States', *Journal of Public Health Dentistry* 56 (5), pp. 253–8.

5. Knox, E. G. (1985), *Fluoridation of Water and Cancer: A Review of the Epidemiological Evidence. Report of the Working Party* (London: HMSO).

6. McDonagh, M. S., et al. (2000), 'Systematic review of water fluoridation', *British Medical Journal* 321 (7265), pp. 855–9.

7. Murray, J. J. (1993), 'Efficacy of preventive agents for dental caries. Systemic fluorides: water fluoridation', *Caries Research* 27 (suppl. 1), pp. 2–8.

8. National Research Council (US) Subcommittee on Health Effects of Ingested Fluoride, Committee on Toxicology, Board on Environmental Studies and Toxicology, Commission on Life Sciences (1993), *Health Effects of Ingested Fluoride* (Washington, D.C.: National Academy Press).

9. Newbrun, E. (1989), 'Effectiveness of water fluoridation', *Journal of Public Health Dentistry* 49 (5), pp. 279–89.

10. Ripa, L. W. (1993), 'A half-century of community water fluoridation in the United States: review and commentary', *Journal of Public Health Dentistry* 53 (1), pp. 17–44.

11. 'Ten great public health achievements – United States, 1900–1999' (1999), *Morbidity and Mortality Weekly Report* 48 (12), pp. 241–3.

It is safe for babies to sleep in bed with their parents

1. American Academy of Pediatrics, Task Force on Infant Sleep Position and Sudden Infant Death Syndrome (2000), 'Changing

concepts of sudden infant death syndrome: implications for infant sleeping environment and sleep position', *Pediatrics* 105, pp. 650–56.

2. Goode, E. (30 September 1999), 'Baby in parents' bed in danger? US says yes, but others demur', *New York Times*.

3. Hauck, F. R., and J. S. Kemp (1998), 'Bedsharing promotes breast-feeding and AAP Task Force on Infant Positioning and SIDS', *Pediatrics* 102 (3), pp. 662–4.

4. Moon, R. Y., and R. Omron (2002), 'Determinants of infant sleep position in an urban population', *Clinical Pediatrics (Philadelphia)* 41 (8), pp. 569–73.

5. Mosko, S., et al. (1996), 'Infant sleep architecture during bedsharing and possible implications for SIDS', *Sleep* 19 (9), pp. 677–84.

6. National Institute of Child Health and Human Development (US) (2006), 'Safe sleep for your baby: ten ways to reduce the risk of Sudden Infant Death Syndrome (SIDS) (General Outreach)', http://www.nichd.nih.gov/publications/pubs/safe_sleep_gen.cfm (accessed 20 May 2008).

7. Person, T. L., W. A. Lavezzi and B. C. Wolf (2002), 'Cosleeping and sudden unexpected death in infancy', *Archives of Pathology and Laboratory Medicine* 126 (3), pp. 343–5.

8. Scheers, N. J., G. W. Rutherford and J. S. Kemp (2003), 'Where should infants sleep? A comparison of risk for suffocation of infants sleeping in cribs, adult beds, and other sleeping locations', *Pediatrics* 112 (4), pp. 883–9.

9. UNICEF UK Baby Friendly Initiative (2008), 'Sharing a bed with your baby: a guide for breastfeeding mothers', http://www.baby friendly.org.uk/pdfs/sharingbedleaflet.pdf (accessed 9 January 2009).

More people commit suicide around public holidays

1. Ambade, V. N., H. V. Godbole and H. G. Kukde (2007), 'Suicidal and homicidal deaths: a comparative and circumstantial approach', *Journal of Forensic and Legal Medicine* 14 (5), pp. 253–60.

2. Annenberg Public Policy Center, University of Pennsylvania (2001), 'Media continue to perpetuate myth of winter holiday–suicide link', http://www.annenbergpublicpolicycenter.org/Downloads/ Adolescent_Risk/Suicide/myth_holiday_suicides20011204.PDF (accessed 2 June 2008).

3. Annenberg Public Policy Center, University of Pennsylvania (2007), 'Holiday–suicide link: newspapers turn the corner', http:// www.annenbergpublicpolicycenter.org/Downloads/Releases/ Release_HolidaySuicide_111907/suicidereleasenov152007final. pdf (accessed 2 June 2008).

4. Bridges, F. S. (2004), 'Rates of homicide and suicide on major national holidays', *Psychological Reports* 94 (2), pp. 723–4.

5. Centers for Disease Control and Prevention (US) (2006), 'Understanding suicide fact sheet', http://www.cdc.gov/ncipc/pub-res/ Suicide%20Fact%20Sheet.pdf (accessed 2 June 2008).

6. Christoffel, K. K., et al. (1988), 'Adolescent suicide and suicide attempts: a population study', *Pediatric Emergency Care* 4 (1), pp. 32–40.

7. Corcoran, P., et al. (2004), 'Temporal variation in Irish suicide rates', *Suicide and Life Threatening Behavior* 34 (4), pp. 429–38.

8. Doshi, A., et al. (2005), 'National study of US emergency department visits for attempted suicide and self-inflicted injury, 1997–2001', *Annals of Emergency Medicine* 46 (4), pp. 369–75.

9. Durkheim, Émile (2006), *On Suicide* (London and New York: Penguin Books).

10. Hillard, J. R., J. M. Holland and D. Ramm (1981), 'Christmas and psychopathology. Data from a psychiatric emergency room population', *Archives of General Psychiatry* 38 (12), pp. 1377–81.

11. Nishi, M., et al. (2000), 'Relationship between suicide and holidays', *Journal of Epidemiology* 10 (5), pp. 317–20.

12. Panser, L. A., et al. (1995), 'Timing of completed suicides among residents of Olmsted County, Minnesota, 1951–1985', *Acta Psychiatrica Scandinavica* 92 (3), pp. 214–19.

13. Valtonen, H., et al. (2006), 'Time patterns of attempted suicide', *Journal of Affective Disorders* 90 (2–3), pp. 201–7.

14. Zonda, T., K. Bozsonyi and E. Veres (2005), 'Seasonal fluctuation of suicide in Hungary, 1970–2000', *Archives of Suicide Research* 9 (1), pp. 77–85.

Newer drugs are always better

1. Agranat, I., H. Caner and J. Caldwell (2002), 'Putting chirality to work: the strategy of chiral switches', *Nature Reviews Drug Discovery* 1 (10), pp. 753–68.
2. ALLHAT (2002), 'Major outcomes in high-risk hypertensive patients randomized to angiotensin-converting enzyme inhibitor or calcium channel blocker vs diuretic: The Antihypertensive and Lipid-Lowering Treatment to Prevent Heart Attack Trial (ALLHAT)', *Journal of the American Medical Association* 288 (23), pp. 2981–97.
3. Angell, Marcia (2004), *The Truth About the Drug Companies: How They Deceive Us and What To Do About It* (New York: Random House).
4. Anonymous (2003), 'Levocetirizine: new preparation. Me-too: simply the active enantiomer of cetirizine', *Prescrire International* 12 (67), pp. 171–2.
5. Wahlqvist, P., et al. (2002), 'Cost effectiveness of esomeprazole compared with omeprazole in the acute treatment of patients with reflux oesophagitis in the UK', *Pharmacoeconomics* 20 (4), pp. 279–87.

Vaccines cause autism

1. BBC News (27 March 2008), 'MMR doctor defends his research', http://news.bbc.co.uk/1/hi/health/7314144.stm (accessed 28 May 2008).
2. Dales, L., S. J. Hammer and N. J. Smith (2001), 'Time trends in

autism and in MMR immunization coverage in California', *Journal of the American Medical Association* 285 (9), pp. 1183–5.

3. Deer, Brian (8 February 2009), 'MMR doctor Andrew Wakefield fixed data on autism', *Sunday Times*, http://www.timesonline.co.uk/tol/life_and_style/health/article5683671.ece (accessed 11 February 2009).

4. Demicheli, V., et al. (2005), 'Vaccines for measles, mumps and rubella in children', *Cochrane Database of Systematic Reviews* (4), CD004407.

5. Madsen, K. M., et al. (2002), 'A population-based study of measles, mumps, and rubella vaccination and autism', *New England Journal of Medicine* 347 (19), pp. 1477–82.

6. National Public Radio website (US) (7 March 2008), 'Case stokes debate about autism, vaccines', http://www.npr.org/templates/story/story.php?storyId=87974932 (accessed 28 May 2008).

7. Nield, L. S. (2006), 'Update on the MMR–autism debate: no evidence for a causative link', *Consultant for Pediatricians* 5, pp. S13–17.

8. Offit, Paul A. (2008), 'Vaccines and autism revisited – the Hannah Poling case', *New England Journal of Medicine* 358 (20), pp. 2089–91.

9. Taylor, B., et al. (1999), 'Autism and measles, mumps, and rubella vaccine: no epidemiological evidence for a causal association', *Lancet* 353 (9169), pp. 2026–9.

10. Wakefield, A. J., et al. (1998), 'Ileal-lymphoid-nodular hyperplasia, non-specific colitis, and pervasive developmental disorder in children', *Lancet* 351 (9103), pp. 637–41.

11. World Health Organization (2006), 'Vaccine-preventable diseases', http://www.who.int/mediacentre/events/2006/g8summit/vaccines/en/index.html (accessed 23 September 2008).

Acknowledgements

No one can write a book alone, and so we would like to thank everyone who helped us to make this a reality. We would like to thank the Indiana University School of Medicine Department of Pediatrics and the Division of Children's Health Services Research, as well as Riley Hospital for Children for their support. We would especially like to thank Dr Richard Schreiner and Dr Stephen Downs, who allow us the freedom to do work off the beaten track. We also thank the Regenstrief Institute and Tom Inui, who has offered us both mentorship and encouragement. We would like to thank our agents, the witty know-it-all Janet Rosen, Sheree Bykofsky, who literally wrote the book which helped us get our foot in their door, and their intern Nathan Belofsky, who picked our query out of the pile. Alyse Diamond, our editor, took over this book at the last second but loved us just the same, and for that we are grateful as well. We would also like to thank the *British Medical Journal* for publishing our myth papers, Toni for her help with hymens, Cindy for leading us through the media maze and Zarafina for producing the best darn tea maker in the world.

Aaron would additionally like to thank his family – his parents (Shelley and Stan), for their lifetime of love and support; his brother and sister-in-law (David and Lisa), for their wisdom and sense of humour; his sister (Nancy), for always being so proud; and his in-laws (Michael, Sharon, Daniel, Mark and Julie), who loved him right from the start. He would also like to thank his friends, who act like family, including (but not limited to) Jon, Sue, Todd, Marlo, Rob, Amy, Gabe, Michelle, Matt, Jill, and especially Todd and Linda – all of whom shamelessly promote this book whenever they get the chance. Aaron is especially grateful for his children: Jacob, who seems to be made only of the best parts of him; Noah, who (unlike Aaron) is fearless and always the life of the party; and Sydney, who completed the family and has promised him she will always be Daddy's little girl. Most of all, he couldn't live without his wife, Aimee, whom he loves dearly, and who loves him enough to let him think he's right most of the time.

Rachel owes many, many thanks to Joe Fick, who makes her laugh and makes her dinner and makes her life full of love. She would also like to thank her parents (Tom and Jacki), who offered her a foundation of love from which to explore the world; her brothers (Dan and Phil), who discovered science with her before any of them realized that scientists don't usually wear disguises; her brothers-in-law (Dave and Charles), who educate her in the ways of television and film quotation; her parents-in-law (Gary and Mae Ellen), who gave her Joe, CFO and a new understanding of love; and all of the Vreemans for whom laughter, books and arguing your point are among life's great pleasures. In addition, Rachel gives her sincere thanks to her beloved circle of sisters in the US (Jessica, Lorrie, Maria, Elizabeth, Martha and Jessica), with special thanks for making her an auntie. She thanks the Fountain Square Supper Club, which indeed makes it a Pleasant street. And she thanks all of the permanent and itinerant occupants of the IU House who make up her family in Eldoret, Kenya, with special gratitude for the constant inspiration from Joe and Sarah Ellen Mamlin; the constant cheering on from Sonak, Karin, April and Martin; and the constant hard work from all of her AMPATH colleagues and friends. Finally, Rachel would like to acknowledge that Aaron was right about this.

Author Biographies

Aaron Carroll, MD, MS, is an Associate Professor of Pediatrics and Director of the Center for Health Policy and Professionalism Research at the Indiana University School of Medicine. He has earned a BA in Chemistry from Amherst College, an MD from the University of Pennsylvania School of Medicine, and a Masters degree in Health Services from the University of Washington. He is a collector of comic books, a whizz with all things technological and a really good video-game player. Still, he managed to get married, and lives with his wife and three children in Carmel, Indiana.

Rachel Vreeman, MD, MS, is an Assistant Professor of Pediatrics in the Children's Health Services Research Group at the Indiana University School of Medicine and Co-Director of Pediatric Research for the Academic Model for the Prevention and Treatment of HIV/AIDS (AMPATH). She has earned a BA in English from Cornell University, an MD from the Michigan State University College of Human Medicine, and a Masters degree in Clinical Research from Indiana University. She enjoys photography, playing the piano, spending time with her husband and letting Aaron know when he is wrong. She divides her time between Indianapolis, Indiana, and Eldoret, Kenya.

Index